Algebra 1

LARSON
BOSWELL
KANOLD
STIFF

Applications • Equations • Graphs

Basic Skills Workbook: Diagnosis and Remediation Teacher's Edition

The Basic Skills Workbook provides material you can use to review and practice basic prerequisite skills for Algebra 1. A diagnostic test is provided to help you determine which topics in the workbook need to be reviewed. The Teacher's Edition includes the student workbook plus assessment materials and answers.

McDougal Littell
A HOUGHTON MIFFLIN COMPANY
Evanston, Illinois • Boston • Dallas

A Note to Teachers

The Student Workbook includes about a month of instruction on topics that are prerequisites for Algebra 1. Each of the five topics comprises four brief lessons with exercises, and each lesson is accompanied by a separate sheet of Quick Check exercises that review the previous lesson and a sheet of extra practice to follow the lesson.

To help you determine which topics will benefit your students, a diagnostic test is provided at the beginning of the Student Workbook. Individual items on the diagnostic test are keyed to each lesson.

The Teacher's Edition also includes a brief assessment following each topic, a cumulative assessment following the last topic, and a complete set of answers.

ISBN: 0-618-02051-9

56789-CKI- 04 03 02

Contents

NAME _____ DATE _____

Diagnostic Test

For use before Topic 1

Factors and Multiples (Topic 1, Lesson 1, pages 1–5)

List all the factors of each number. Name the greatest common factor and the least common multiple of the two numbers.

1. 21, 49   **2.** 99, 33   **3.** 48, 52

Write the prime factorization of the number. If the number is prime, write *prime*.

4. 84   **5.** 117   **6.** 41

Comparing and Ordering Numbers (Topic 1, Lesson 2, pages 6–10)

Compare the two numbers. Write your answer using <, >, or =.

7. 77 ◯ 71   **8.** 111.10 ◯ 111.08

9. $2\frac{4}{5}$ ◯ $\frac{14}{5}$

Write the numbers in order from least to greatest.

10. 3368, 3168, 3367, 3370   **11.** 16.01, 16.005, 16.42, 16.0009

12. $\frac{1}{2}, \frac{3}{7}, \frac{2}{3}, \frac{8}{9}$ **13.** $3\frac{1}{2}, 4\frac{1}{3}, 4\frac{1}{2}, 3\frac{3}{4}$

Whole Number and Decimal Operations
(Topic 1, Lesson 3, pages 11–16)

Find the sum or difference.

14. $169 + 215$ **15.** $368.5 - 79.83$

16. $72.62 + 84.9 - 56.48$

Find the product or quotient.

17. 8.2×4.3 **18.** $5.35 \div 0.5$ **19.** $96 \div 6.25$

Fraction Operations (Topic 1, Lesson 4, pages 17–22)

Find the sum or difference.

20. $\frac{1}{3} + \frac{7}{9}$ **21.** $\frac{3}{10} - \frac{1}{5}$ **22.** $8\frac{5}{8} - 3\frac{23}{24}$

1. Factors: _____
   GCF: _____
   LCM: _____
2. Factors: _____
   GCF: _____
   LCM: _____
3. Factors: _____
   GCF: _____
   LCM: _____
4. _____
5. _____
6. _____
7. _____
8. _____
9. _____
10. _____
11. _____
12. _____
13. _____
14. _____
15. _____
16. _____
17. _____
18. _____
19. _____
20. _____
21. _____
22. _____

NAME _____ DATE _____

Diagnostic Test

For use before Topic 1

Find the reciprocal of the number.

23. $\frac{3}{5}$ 24. 16   25. $6\frac{3}{7}$

Find the product or quotient.

26. $\frac{2}{4} \times 18$ 27. $\frac{1}{6} \div \frac{9}{14}$ 28. $3\frac{5}{8} \div \frac{11}{12}$

Mean, Median, Mode, and Range (Topic 2, Lesson 1, pages 24–28)

Find the mean, the median, the mode(s), and the range for each set of data. If necessary, round your anwers to the nearest hundredth.

29. Cost of CDs: $11.99, $8.95, $12.99, $14.95, $12.99

30.

All-Time Leading Touchdown Scorers	
Jerry Rice	165
Marcus Allen	134
Jim Brown	126
Walter Payton	125
John Riggins	116

Bar Graphs and Line Graphs (Topic 2, Lesson 2, pages 29–34)

For Exercises 31 and 32, use the bar graph.

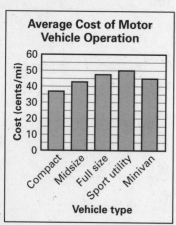

Average Cost of Motor Vehicle Operation

31. Estimate the average cost of operating a sport utility vehicle.

32. Which types of vehicles cost 45 cents/mile or less?

33. Draw a line graph to display the data in the table.

Average High Temperature (F°) in Portland, Oregon											
Jan.	Feb.	Mar.	Apr.	May	June	July	Aug.	Sept.	Oct.	Nov.	Dec.
45°	51°	56°	61°	67°	74°	80°	80°	75°	64°	53°	46°

23. _____
24. _____
25. _____
26. _____
27. _____
28. _____
29. Mean: _____
    Median: _____
    Mode: _____
    Range: _____
30. Mean: _____
    Median: _____
    Mode: _____
    Range: _____
31. _____
32. _____
33. _____

Diagnostic Test

For use before Topic 1

Circle Graphs (Topic 2, Lesson 3, pages 35–39)

For Exercises 34 and 35, use the circle graph.

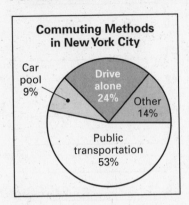

Commuting Methods in New York City

Car pool 9%
Drive alone 24%
Other 14%
Public transportation 53%

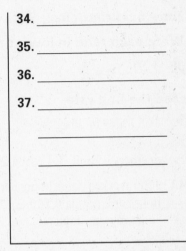

34. _____

35. _____

36. _____

37. _____

_____

_____

_____

_____

_____

34. Suppose the number of people commuting in New York City is about 9 million. About how many people take public transportation?

35. Suppose the number of people commuting in New York City is about 9 million. About how many more people drive alone than carpool?

Interpreting Graphs (Topic 2, Lesson 4, pages 40–44)

For Exercises 36 and 37, use the two graphs below that illustrate the same data.

Graph A

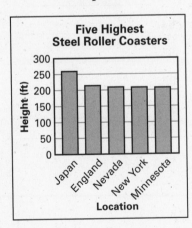

Five Highest Steel Roller Coasters

Height (ft)

Japan England Nevada New York Minnesota

Location

Graph B

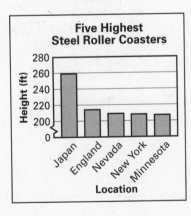

Five Highest Steel Roller Coasters

Height (ft)

Japan England Nevada New York Minnesota

Location

36. Which graph suggests that the Roller Coaster in Japan is more than twice as high as the Roller Coaster in England? Is this impression correct?

37. Why do these graphs give such a different visual impression?

Name _____ Date _____

Diagnostic Test
For use before Topic 1

Rates and Ratios (Topic 3, Lesson 1, pages 46–50)

Write each ratio in lowest terms.

38. 5:10

39. 18:15

Find the unit rate.

40. 165 miles in 3 hours

41. $18,000 in 12 months

Equal Rates (Topic 3, Lesson 2, pages 51–54)

42. John worked for 6 hours and got paid $31.50. Rhonda worked for 9 hours and got paid $46.80. Are both John and Rhonda paid the same rate per hour?

Find the missing number.

43. $\frac{9}{12} = \frac{?}{4}$

44. $\frac{5}{12} = \frac{15}{?}$

Fractions, Decimals, and Percents (Topic 3, Lesson 3, pages 55–60)

Write each fraction or mixed number as a decimal.

45. $\frac{9}{10}$

46. $1\frac{5}{8}$

47. $2\frac{2}{3}$

Write each fraction or mixed number in lowest terms.

48. $\frac{12}{36}$

49. $6\frac{8}{24}$

50. $\frac{28}{49}$

Write each percent as a decimal and as a fraction or mixed number in lowest terms.

51. $66.\overline{6}\%$

52. 0.2%

53. 150%

Finding a Percent of a Number (Topic 3, Lesson 4, pages 61–64)

54. What number is 40% of 25?

55. 10% of 300 is what number?

56. What number is 50% of 96?

57. 22% of 45 is what number?

38. _____
39. _____
40. _____
41. _____
42. _____
43. _____
44. _____
45. _____
46. _____
47. _____
48. _____
49. _____
50. _____
51. _____
52. _____
53. _____
54. _____
55. _____
56. _____
57. _____

Algebra 1
Basic Skills: Diagnosis and Remediation

NAME _____ DATE _____

Diagnostic Test

For use before Topic 1

Patterns in Geometry   (Topic 4, Lesson 1, pages 66–69)

For Exercises 58–59, use the figures to find a pattern.

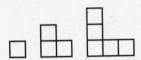

58. Draw the next two figures in the pattern.

59. Use the table to predict how many squares will be in the seventh figure in the pattern.

Figure Number	1	2	3
Number of squares	1	3	5

Polygons   (Topic 4, Lesson 2, pages 70–74)

Complete the statement.

60. A _____ is where the sides of a polygon meet.

61. A polygon with four sides is a _____.

62. A rectangle that has four congruent sides is a _____.

63. Sketch a triangle that is not regular.

Perimeters and Areas of Polygons   (Topic 4, Lesson 3, pages 75–79)

Find the perimeter and area of each polygon.

64.

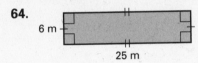

6 m

25 m

65.

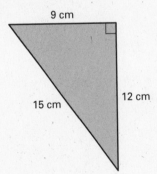

9 cm

15 cm

12 cm

Circles and Areas   (Topic 4, Lesson 4, pages 80–84)

66. Find the circumference and area of a circle with diameter of 36 ft. Use 3.14 for π.

58. _____

59. _____

60. _____

61. _____

62. _____

63.

64. Perimeter: _____

   Area: _____

65. Perimeter: _____

   Area: _____

66. Circumference: _____

   Area: _____

NAME _____ DATE _____

Diagnostic Test

For use before Topic 1

Integer Concepts (Topic 5, Lesson 1, pages 87–91)

Find the opposite and the absolute value of the integer.

67. 16   **68.** -8

Write the set of integers in order from least to greatest.

69. 1, 0, -3 **70.** 6, -2, 5, -7

Adding Integers (Topic 5, Lesson 2, pages 92–95)

Find each sum.

71. $-12 + 15$ **72.** $-25 + (-41)$ **73.** $-24 + 21$

Subtracting Integers (Topic 5, Lesson 3, pages 96–100)

Find each difference.

74. $-15 - 30$ **75.** $-28 - (-20)$ **76.** $24 - 40$

Multiplying and Dividing Integers (Topic 5, Lesson 4, pages 101–105)

Find each product or quotient.

77. $8(-5)$ **78.** $-48 \div (-6)$ **79.** $(-3)(7)(-2)$

67. Opposite: _____

    Absolute Value: _____

68. Opposite: _____

    Absolute Value: _____

69. _____

70. _____

71. _____

72. _____

73. _____

74. _____

75. _____

76. _____

77. _____

78. _____

79. _____

NAME _____ DATE _____

Factors and Multiples

GOAL Find the prime factorization, the least common multiple, and the greatest common factor of given numbers.

Knowing how to write a natural number as the product of prime numbers can help you find a least common denominator when adding and subtracting fractions. Below are some important characteristics of prime numbers.

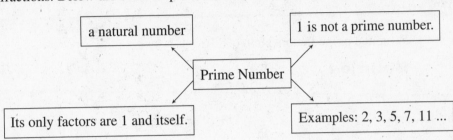

| a natural number | | 1 is not a prime number. |

Prime Number

| Its only factors are 1 and itself. | | Examples: 2, 3, 5, 7, 11 ... |

Terms to Know

Terms to Know	Example/Illustration
Natural Numbers all the numbers in the sequence 1, 2, 3, 4, 5, . . .	1 2 8 10 32
Factor a number that is multiplied in an expression	In the equation 2 · 4 = 8, 2 and 4 are factors.
Prime Number a natural number that has exactly two factors, itself and 1	The factors of 19 are 19 and 1, so 19 is a prime number.
Prime Factorization number written as the product of prime numbers	$24 = 2 \cdot 2 \cdot 2 \cdot 3 = 2^3 \cdot 3$
Common Factor number that divides two or more given natural numbers evenly	7 is a common factor of 21 and 35 because 7 · 3 = 21 and 7 · 5 = 35.
Greatest Common Factor the largest number that is a common factor of two or more natural numbers	6 is the greatest common factor of 18 and 24 because 6 · 3 = 18 and 6 · 4 = 24.
Multiple A multiple of a natural number is the product of the given number and any natural number.	6 and 12 are multiples of 3 because 3 · 2 = 6 and 3 · 4 = 12.
Least Common Multiple the smallest number that is a multiple of two or more given natural numbers	The least common multiple of 2 and 3 is 6.

(continued)

NAME _____ DATE _____

Factors and Multiples

Understanding the Main Ideas

You can draw tree diagrams to write the prime factorization of a number as shown in the following example.

EXAMPLE 1 _____

Write the prime factorization of 126.

SOLUTION

$$
\begin{array}{c}
126 \\
2 \times 63 \\
2 \times 3 \times 21 \\
2 \times 3 \times 3 \times 7
\end{array}
$$

The prime factorization of 126 is $2 \cdot 3^2 \cdot 7$.

Write the prime factorization.

1. 18   **2.** 66   **3.** 468   **4.** 350

When finding the greatest common factor, follow these steps.

1. Write the prime factorization of each number.

2. Multiply the common prime factors to find the greatest common factor.

EXAMPLE 2 _____

Find the greatest common factor of 40 and 72.

SOLUTION

Step 1: Write the prime factorization of each number.

$40 = 2 \cdot 2 \cdot 2 \cdot 5$

$72 = 2 \cdot 2 \cdot 2 \cdot 3 \cdot 3$

Step 2: Multiply the common factors.

$2 \cdot 2 \cdot 2 = 8$

The greatest common factor is 8.

Find the greatest common factor.

5. 27, 45   **6.** 75, 27   **7.** 52, 36, 132   **8.** 19, 41, 53

(continued)

NAME _____ DATE _____

Factors and Multiples

When finding the least common multiple, follow these steps.

1. List the prime factorization of the numbers.

2. Circle any factors the numbers have in common.

3. List the factors of both, writing any circled factors only once.

4. Multiply the factors in the last list to get the least common multiple.

EXAMPLE 3 _____

Find the least common multiple of 28 and 32.

SOLUTION

Step 1: Write the prime factorization of each number.

$28 = 2 \cdot 2 \cdot 7$

$32 = 2 \cdot 2 \cdot 2 \cdot 2 \cdot 2$

Step 2: Circle any common factors.

$28 = \boxed{2} \cdot \boxed{2} \cdot 7$

$32 = \boxed{2} \cdot \boxed{2} \cdot 2 \cdot 2 \cdot 2$

Step 3: List the factors of both, writing the circled ones only once.

2, 2, 7, 2, 2, 2

Step 4: Multiply the factors in the list.

$2 \times 2 \times 7 \times 2 \times 2 \times 2 = 224$

The least common multiple is 224.

Find the least common multiple.

9. 35, 40 **10.** 64, 72 **11.** 46, 115 **12.** 2, 3, 7

Mixed Review

13. Estimate the sum $12.2 + 15.7 + 19.8$.

14. Complete: $3.6 \text{ km} = $ ____ m.

Algebra 1
Basic Skills Workbook: Diagnosis and Remediation

NAME _____ DATE _____

Quick Check

Review geometry, writing expressions, rounding, and checking solutions

Standardized Testing Quick Check

1. In the figure at the right, what is the total number of lines?

   A. 3 lines

   B. 6 lines

   C. 9 lines

   D. 12 lines

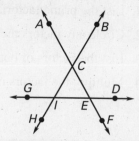

2. Anna and Carl Simko plan to take their three children to a movie. The ticket prices are $7 for adults and $4.75 for children. They will also spend $15 on refreshments and $5 on parking. Which expression could the Simkos use to find their total cost?

   A. $2(\$7) + 3(\$4.75) + \$20$

   B. $\$7 + \$4.75 + \$15 + \$5$

   C. $3(\$7) + 2(\$4.75) + \$5 + \$15$

   D. $2(\$7 + \$15) + 3(\$4.75 + \$5)$

   E. None of these

Homework Review Quick Check

3. Round 6.235 to the nearest tenth.

4. Is 3 a solution of $3n - 1 = 10$?

NAME _____ DATE _____

Practice

For use with Lesson 1.1: Factors and Multiples

List all the factors of the number.

1. 15   **2.** 26   **3.** 99   **4.** 35

5. 144   **6.** 49   **7.** 61   **8.** 56

9. 72   **10.** 98   **11.** 63   **12.** 196

Write the prime factorization of the number. If a number is prime, write *prime*.

13. 16   **14.** 17   **15.** 9   **16.** 27

17. 20   **18.** 42   **19.** 81   **20.** 256

21. 65   **22.** 73   **23.** 55   **24.** 48

25. 120   **26.** 252   **27.** 133   **28.** 101

List all the common factors of the pair of numbers.

29. 25, 14   **30.** 40, 72   **31.** 6, 54   **32.** 12, 24

33. 35, 36   **34.** 51, 17   **35.** 8, 56   **36.** 45, 18

Find the greatest common factor of the pair of numbers.

37. 12, 15   **38.** 17, 85   **39.** 48, 56   **40.** 42, 72

41. 35, 36   **42.** 11, 23   **43.** 26, 34   **44.** 57, 102

45. 13, 120   **46.** 64, 125   **47.** 104, 22   **48.** 150, 240

49. 300, 550   **50.** 458, 310   **51.** 6, 8, 12   **52.** 14, 63, 84

Find the least common multiple of the pair of numbers.

53. 4, 12   **54.** 24, 18   **55.** 20, 25   **56.** 9, 15

57. 14, 21   **58.** 13, 16   **59.** 28, 32   **60.** 23, 56

61. 17, 51   **62.** 64, 32   **63.** 72, 144   **64.** 48, 60

65. 75, 100   **66.** 15, 25, 75 **67.** 5, 6, 10   **68.** 22, 55, 60

NAME _____ DATE _____

Comparing and Ordering Numbers

GOAL **Compare and order numbers.**

> Matt ran in a 5K race and wanted to know how fast he ran the race compared with his five friends. Matt finished the race in 20.54 minutes. The times of his friends were 19.94, 20.23, 21.46, 20.97, and 22.12 minutes. You can compare and order their times to find out who ran the race the fastest.

Terms to Know

Example/Illustration

Least Common Denominator (LCD) least common multiple of the denominators	The least common denominator of $\frac{5}{6}$ and $\frac{3}{4}$ is 12 because the least common multiple of 6 and 4 is 12.

Understanding the Main Ideas

When you compare two numbers, *a* and *b*, there are exactly three possibilities.

a is less than b.	$a < b$
a is equal to b.	$a = b$
a is greater than b.	$a > b$

To compare two whole numbers or decimals, compare the digits of the two numbers from left to right. Find the first place in which the digits are different.

EXAMPLE 1

Compare the two numbers 6589 and 6598. Write the answer using <, =, or >.

SOLUTION

6 5 ⑧ 9   Compare the digits from left to right.

6 5 ⑨ 8   Find the first place where the digits are different.

8 < 9, so 6589 < 6598

You can picture this on a number line. The numbers on a number line increase from left to right.

6589 is *less* than 6598.

6589 is to the *left* of 6598.

Compare the numbers. Write the answer using <, =, or >.

1. 10,580 and 8599   **2.** 875 and 873   **3.** 165,223 and 165,142

(continued)

Comparing and Ordering Numbers

EXAMPLE 2

Compare the two numbers 34.23 and 34.55. Write the answer using <, =, or >.

SOLUTION

3 4 . ② 3   Compare the digits from left to right.

3 4 . ⑤ 5   Find the first place where the digits are different.

5 > 2, so 34.55 > 34.23

You can picture this on a number line. The numbers on a number line increase from left to right.

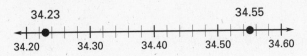

34.55 is *greater* than 34.23

34.55 is to the *right* of 34.23.

Compare the numbers. Write the answer using <, =, or >.

4. 18.57 and 21.57   **5.** 0.0039 and 0.027   **6.** 65.81 and 65.801

To compare two fractions that have the same denominator, compare the numerators. If the fractions have different denominators, first rewrite one or both fractions to produce equivalent fractions with a common denominator.

EXAMPLE 3

Write the numbers $\frac{2}{3}$, $\frac{7}{15}$, and $\frac{1}{2}$ in order from least to greatest.

SOLUTION

The LCD of the fractions is 30.

$$\frac{2}{3} = \frac{2 \cdot 10}{3 \cdot 10} = \frac{20}{30} \qquad \frac{7}{15} = \frac{7 \cdot 2}{15 \cdot 2} = \frac{14}{30} \qquad \frac{1}{2} = \frac{1 \cdot 15}{2 \cdot 15} = \frac{15}{30}$$

Compare the numerators: $14 < 15 < 20$, so $\frac{7}{15} < \frac{1}{2} < \frac{2}{3}$.

In order from least to greatest, the fractions are $\frac{7}{15}$, $\frac{1}{2}$, and $\frac{2}{3}$.

Write the numbers in order from least to greatest.

7. 1304, 1430, 1340, 1334   **8.** 3.248, 3.284, 3.481, 3.847

9. $\frac{1}{3}, \frac{5}{6}, \frac{5}{8}, \frac{6}{5}$ **10.** $\frac{3}{8}, \frac{3}{4}, \frac{1}{3}, \frac{7}{12}$

(continued)

Algebra 1
Basic Skills Workbook: Diagnosis and Remediation

7

Topic 1

NAME _____ DATE _____

Comparing and Ordering Numbers

EXAMPLE 4

Compare $2\frac{3}{5}$ and $2\frac{2}{3}$. Write the answer using <, =, or >.

SOLUTION

The whole number parts of the mixed numbers are the same, so compare the fraction parts.

The LCD of $\frac{3}{5}$ and $\frac{2}{3}$ is 15.

$$\frac{3}{5} = \frac{3 \cdot 3}{5 \cdot 3} = \frac{9}{15} \qquad \frac{2}{3} = \frac{2 \cdot 5}{3 \cdot 5} = \frac{10}{15}$$

Compare the numerators: $9 < 10$, so $\frac{3}{5} < \frac{2}{3}$.

Because $\frac{3}{5} < \frac{2}{3}$, it follows that $2\frac{3}{5} < 2\frac{2}{3}$.

$2\frac{3}{5}$ is to the left of $2\frac{2}{3}$.

Compare the two numbers. Write the answer using <, =, or >.

11. $19\frac{7}{9}$ and $19\frac{1}{4}$ **12.** $\frac{17}{3}$ and $5\frac{2}{3}$ **13.** $8\frac{1}{6}$ and $8\frac{7}{9}$

Mixed Review

Find the greatest common factor.

14. 18, 42 **15.** 64, 40

16. 121, 143 **17.** 13, 97

Find the least common multiple.

18. 7, 9 **19.** 70, 28

20. 85, 125 **21.** 12, 4, 24

Algebra 1
Basic Skills Workbook: Diagnosis and Remediation

Topic 1

NAME _____ DATE _____

Quick Check

Review of Topic 1, Lesson 1

Standardized Testing Quick Check

1. Jeffrey has 104 trading cards. David has 32 trading cards. Each package that they bought had the same number of cards. What is the largest possible number of cards that could come in each package?

   A. 4 cards

   B. 8 cards

   C. 13 cards

   D. 21 cards

2. On average, your car can go 300 miles before you have to fill the tank with gas. It can go 3000 miles before a recommended oil change. After how many miles will both the gas tank need to be filled and the oil need to be changed?

   A. 300 miles

   B. 900,000 miles

   C. 90,000 miles

   D. 3000 miles

Homework Review Quick Check

Find the greatest common factor.

3. 48, 54   **4.** 16, 40   **5.** 5, 7, 13

Find the least common multiple.

6. 21, 49   **7.** 19, 51   **8.** 3, 9, 12

NAME _____ DATE _____

Practice

For use with Lesson 1.2: Comparing and Ordering Numbers

Compare the two numbers. Write the answer using <, =, or >.

1. 12,868 and 14,653   **2.** 643 and 623   **3.** 26,555 and 26,653

4. 24.53 and 26.98   **5.** 0.00652 and 0.6052   **6.** 84.35 and 84.3

7. 6003.7 and 6307.04   **8.** 571,364 and 571,377   **9.** 0.168 and 0.0085

10. $\frac{3}{8}$ and $\frac{5}{8}$ **11.** $2\frac{4}{7}$ and $\frac{18}{7}$ **12.** $\frac{4}{5}$ and $\frac{3}{4}$

13. $\frac{8}{36}$ and $\frac{2}{9}$ **14.** $\frac{5}{9}$ and $\frac{1}{2}$ **15.** $6\frac{5}{13}$ and $5\frac{3}{5}$

16. $\frac{1}{3}$ and $\frac{1}{4}$ **17.** $5\frac{3}{4}$ and $\frac{23}{4}$ **18.** $\frac{2}{3}$ and $\frac{6}{8}$

19. 75,663 and 75,663.01   **20.** 265.05 and 265.01   **21.** 8978.99 and 8979

Write the numbers in order from least to greatest.

22. 8450, 8054, 5840, 8455   **23.** 23,546, 23,766, 20,599, 21,431

24. 12.52, 21.25, 12.05, 21.257   **25.** 7.0056, 7.059, 7.0074, 7.0008

26. $\frac{2}{3}, \frac{3}{4}, \frac{1}{6}, \frac{7}{8}$ **27.** $\frac{1}{2}, \frac{3}{10}, \frac{1}{4}, \frac{3}{5}, \frac{9}{20}$

28. $7\frac{7}{9}, 7\frac{1}{2}, 8\frac{1}{6}, 7\frac{2}{3}, 8\frac{5}{18}$ **29.** $\frac{9}{7}, 1\frac{2}{5}, 1\frac{1}{10}, \frac{55}{35}$

30. $\frac{11}{9}, \frac{13}{3}, \frac{17}{11}, \frac{15}{9}$ **31.** $14\frac{1}{3}, 15\frac{4}{5}, 14\frac{13}{15}, 15\frac{1}{2}$

32. Glen finished the 5K race in 20.75 minutes. Mark finished the race in 20.36 minutes. Which runner finished the race first?

33. You need $10\frac{2}{3}$ feet of shelving to make shelves for a pantry. The salesperson at the home improvement store gives you $10\frac{1}{2}$ feet of shelving. Did the salesperson give you enough to complete the project?

Algebra 1
Basic Skills Workbook: Diagnosis and Remediation

NAME _____ DATE _____

Whole Number and Decimal Operations

GOAL Add, subtract, multiply, and divide whole numbers and decimals.

Whole numbers and decimal operations, such as addition, can be used to help you find the total cost of your grocery bill. You can use subtraction to help you find the change you expect after paying the bill.

Grocery bill	
Cereal	. . . .$2.59
Yogurt	. . . .$ .65
Milk	. . . . . .$2.69
Total	. . . . .$5.93
Cash	. . . .$10.00
Change	. . .$4.07

EXAMPLE 1

Add or subtract.

a. $158 + 365$

b. $460 - 193$

SOLUTION

a.
```
   1 1
   158
 + 365
```
$523 \rightarrow 8 + 5 = 13$
$\qquad\ 1 + 5 + 6 = 12$
$\qquad\ 1 + 1 + 3 = 5$

b.
```
 3 15 10
   460
 - 193
```
$267 \rightarrow 10 - 3 = 7$
$\qquad\ 15 - 9 = 6$
$\qquad\ 3 - 1 = 2$

Add or subtract.

1. $86 - 39$ **2.** $232 + 598$ **3.** $1587 + 26{,}843$ **4.** $743 - 129$

EXAMPLE 2

Multiply or divide.

a. 67×9

b. $95 \div 8$

SOLUTION

a.
```
    6
   67
  × 9
```
$603 \rightarrow 7 \times 9 = 63$
$\qquad\ 9 \times 6 + 6 = 60$

b.
$$11\tfrac{7}{8}$$
$$8\,\overline{)\,95}$$
$\underline{8} \rightarrow 8 \times 10 = 80$
$15 \rightarrow 95 - 80 = 15$
$\underline{8} \rightarrow 8 \times 1 = 8$
$7 \rightarrow 7$ is the remainder.

(continued)

Algebra 1
Basic Skills Workbook: Diagnosis and Remediation

Topic 1

Multiply or divide.

5. 94×61 **6.** 743×47 **7.** $62 \div 24$ **8.** $956 \div 38$

To add and subtract decimals, you can use the vertical format. When you do this, line up the decimal places, adding zeros as placeholders as needed to help keep the decimal places aligned correctly. The steps are similar to those used for adding and subtracting whole numbers, as shown in Example 3.

EXAMPLE 3
_____

Add or subtract.

a. $3.5 + 0.64 + 7$ **b.** $17.4 - 6.85$

SOLUTION

Write each problem in vertical form. Line up the decimal points.
Use zeros as placeholders.

                                         6 13 10

a. 3.50 **b.** $17.\cancel{4}\cancel{0}$

         0.64 $- 6.85$

    $\underline{+ 7.00}$ $\overline{10.55}$

    11.14

Add or subtract.

9. $6.87 + 7.24$ **10.** $0.3 + 9.06$ **11.** $6.75 - 2.3$ **12.** $12 - 7.652$

Decimal multiplication is similar to multiplication with whole numbers. When multiplying with decimals, you need to know where to put the decimal point. The number of decimal places in the product is equal to the sum of the number of decimal places in the factors.

EXAMPLE 4
_____

Multiply 5.23×6.4.

SOLUTION

Write the problem vertically. You do not need to line up the decimal points. The total number of decimal points in the factors is the number of decimal places in the answer.

(continued)

Whole Number and Decimal Operations

```
    5.23    two decimal places
  × 6.4     one decimal place
   2092
   3138
  33.472    three decimal places
```

Multiply.

13. 2.25×5.61 **14.** 6.643×1.495 **15.** 5.1×0.02 **16.** 0.034×8.802

Whether you are dividing a decimal by a whole number or dividing two decimals, the steps for dividing decimals using long division are the same as the steps for long division with whole numbers. When you do long division with decimals, line up the decimal places in the quotient with the decimal places in the dividend.

EXAMPLE 5

Divide: $118.32 \div 12$.

SOLUTION

```
          9.86
   12 ) 118.32        Line up the decimal places.
        108           9 wholes × 12 = 108 wholes
        10 3          Bring down the 3.
         9 6          8 tenths × 12 = 96 tenths
          72          Bring down the 2.
          72          6 hundredths × 12 = 72 hundredths
           0
```

So, $118.32 \div 12 = 9.86$.

EXAMPLE 6

Divide $0.086 \div 0.3$. If necessary, round to the nearest hundredth.

SOLUTION

Write the problem in long division form.

$$0.3 \overline{)\ 0.086}$$

(continued)

NAME _____ DATE _____

Whole Number and Decimal Operations

Move the decimal points in the divisor and dividend the same number of places until the divisor is a whole number.

$$0.3 \overline{)0.086} \qquad \Rightarrow \qquad 3 \overline{)0.86}$$

Next divide. Write zeros in the dividend to continue the division as necessary. If the division doesn't come out evenly, carry it out to the thousandths place to round to the nearest hundredth.

```
       0.286
   3 ) 0.860   ←Write 0 in dividend.
       6
       26
       24
        20   ←Bring down 0.
        18
         2
```

0.286 rounds to 0.29, so 0.086 ÷ 0.3 is about 0.29.

Divide. If necessary, round to the nearest hundredth.

17. 12.7 ÷ 2   **18.** 12.35 ÷ 0.65   **19.** 7.85 ÷ 0.006   **20.** 204.02 ÷ 5.05

Mixed Review
....

Write the prime factorization. If a number is prime, write *prime*.

21. 96   **22.** 31   **23.** 105   **24.** 91

Write the numbers in order from least to greatest.

25. 5.23, 5.79, 5.25, 6.0   **26.** 8556, 8565, 6585, 8557

27. $\frac{5}{9}, \frac{4}{6}, \frac{5}{3}, \frac{1}{2}$ **28.** 179.009, 179.090, 179.087, 179.0089

(continued)

Algebra 1
Basic Skills Workbook: Diagnosis and Remediation

NAME _____ DATE _____

Quick Check

Review of Topic 1, Lesson 2

Standardized Testing Quick Check

1. Which group of numbers is in order from least to greatest, reading from left to right?

   A. 98.7, 98.07, 90.78, 90.87, 97.08

   B. 3.50, 3.05, 3.12, 3.1, 3.01

   C. 0.602, 0.62, 0.630, 0.633, 0.66

   D. 1.23, 12.4, 1.25, 1.26, 12.7

2. If the numbers are ordered from least to greatest, which number could you use to replace n in the list of numbers 0.23, $\frac{2}{5}$, $\frac{2}{3}$, n, and $\frac{3}{4}$?

   A. 0.68

   B. $\frac{1}{2}$

   C. 0.83

   D. $\frac{200}{116}$

Homework Review Quick Check

Write the numbers in order from least to greatest.

3. 5.4, 7.05, 6.8, 6.75, 7.5, 5.45

4. $\frac{8}{7}$, $\frac{18}{6}$, $\frac{7}{3}$, $\frac{7}{9}$, $\frac{7}{2}$, $\frac{13}{4}$

Compare the two numbers. Write the answer using $<$, $=$, or $>$.

5. $\frac{28}{9}$ and $3\frac{1}{9}$

6. $4\frac{7}{20}$ and 4.35

7. $\frac{14}{3}$ and $\frac{17}{4}$

NAME _____ DATE _____

Practice

For use before Lesson 1.3: Whole Number and Decimal Operations

For Exercises 1–21, find the sum or difference.

1. $49 + 81$ **2.** $112 + 254$ **3.** $74 - 53$

4. $117 - 59$ **5.** $7.92 + 6.5$ **6.** $12.36 + 9$

7. $3.42 - 2.4$ **8.** $18.95 - 4$ **9.** $24 - 5.36$

10. $0.681 + 5.5$ **11.** $28.012 + 94.3$ **12.** $0.88 - 0.39$

13. $5.452 - 2.91$ **14.** $28.3 + 8.624$ **15.** $86 - 77.41$

16. $843 + 33.29$ **17.** $0.00902 - 0.00887$ **18.** $199.9938 + 95.992$

19. $7.56 - 2.019 + 5.451$ **20.** $2.25 + 7.789 - 4.342$ **21.** $11.010 + 5.672 - 8.999$

In Exercises 22–42, find the product or quotient.

22. 6.25×6.5 **23.** 0.26×9.58 **24.** $133.6 \div 8$

25. $39.2 \div 7$ **26.** $2.43 \div 0.03$ **27.** 4.25×12.35

28. $57.3 \div 0.003$ **29.** 0.15×24 **30.** $231.84 \div 12.6$

31. $42.37 \div 1.9$ **32.** 10×57.86 **33.** $2.8 \div 0.7$

34. $8.37 \div 0.27$ **35.** 0.985×2.5 **36.** 7.71×9.44

37. $100.38 \div 21$ **38.** $84.4 \div 0.02$ **39.** 64×3.51

40. 183.62×2.834 **41.** $150.375 \div 80.2$ **42.** $2712.15 \div 35$

43. You bought a shirt for $24.00, a pair of pants for $25.99, and shoes for $12.45. How much did you spend altogether?

44. You give the cashier $55.50 for a grocery bill of $51.47. How much change will you get back?

45. Find the area of the rectangle below.

13.5 cm

15.2 cm

Algebra 1
Basic Skills Workbook: Diagnosis and Remediation

Topic 1

NAME _____ DATE _____

Fraction Operations

GOAL Add, subtract, multiply, and divide fractions and mixed numbers.

> You can use fraction operations when dealing with food recipes. For example, if you only have one measuring cup that holds $\frac{1}{3}$ cup of flour and you need $3\frac{1}{3}$ cups of flour, how many times do you have to fill the measuring cup? You can use division to solve this problem.

Terms to Know	Example/Illustration
Reciprocal two numbers whose product is 1	$\frac{5}{6}$ and $\frac{6}{5}$ are reciprocals because $\frac{5}{6} \times \frac{6}{5} = 1$.

Understanding the Main Ideas

To add or subtract two fractions with the same denominators, add or subtract the numerators.

EXAMPLE 1 _____

Subtract $\frac{8}{12} - \frac{4}{12}$.

SOLUTION

$$\frac{8}{12} - \frac{4}{12} = \frac{8-4}{12} \qquad \text{Subtract numerators.}$$

$$= \frac{4}{12} \qquad \text{Simplify.}$$

$$= \frac{1 \cdot \cancel{4}}{3 \cdot \cancel{4}} \qquad \text{Factor numerator and denominator.}$$

$$= \frac{1}{3} \qquad \text{Simplify fraction to lowest terms.}$$

Add or subtract.

1. $\frac{2}{14} + \frac{5}{14}$ **2.** $\frac{7}{8} - \frac{1}{8}$ **3.** $\frac{19}{21} - \frac{7}{21}$ **4.** $\frac{4}{9} + \frac{8}{9}$

To add or subtract two fractions with different denominators, write equivalent fractions with a common denominator.

(continued)

Algebra 1 **17**
Basic Skills Workbook: Diagnosis and Remediation

NAME _____ DATE _____

Fraction Operations

EXAMPLE 2

Add $\frac{2}{5} + \frac{4}{7}$.

SOLUTION

$$\frac{2}{5} + \frac{4}{7} = \frac{14}{35} + \frac{20}{35} \qquad \text{Use the LCD, 35.}$$

$$= \frac{14 + 20}{35} \qquad \text{Add numerators.}$$

$$= \frac{34}{35} \qquad \text{Simplify.}$$

To add or subtract mixed numbers, you can first rewrite them as fractions.

EXAMPLE 3

Subtract $3\frac{4}{5} - 1\frac{1}{4}$.

SOLUTION

$$3\frac{4}{5} - 1\frac{1}{4} = \frac{19}{5} - \frac{5}{4} \qquad \text{Rewrite mixed numbers as fractions.}$$

$$= \frac{76}{20} - \frac{25}{20} \qquad \text{The LCD is 20.}$$

$$= \frac{76 - 25}{20} \qquad \text{Subtract numerators.}$$

$$= \frac{51}{20}, \text{ or } 2\frac{11}{20} \qquad \text{Simplify.}$$

Add or subtract.

5. $\frac{1}{3} + \frac{3}{8}$ **6.** $\frac{5}{6} - \frac{1}{2}$ **7.** $5\frac{3}{4} + 3\frac{1}{2}$ **8.** $9\frac{5}{6} - 3\frac{1}{2}$

To multiply two fractions, multiply the numerators and multiply the denominators.

(continued)

Algebra 1
Basic Skills Workbook: Diagnosis and Remediation

NAME _____ DATE _____

Fraction Operations

EXAMPLE 4

Multiply $\frac{2}{3} \times \frac{6}{11}$.

SOLUTION

$$\frac{2}{3} \times \frac{6}{11} = \frac{2 \times 6}{3 \times 11}$$ Multiply numerators and denominators.

$$= \frac{12}{33}$$ Simplify.

$$= \frac{3 \cdot 4}{3 \cdot 11}$$ Factor numerator and denominator.

$$= \frac{4}{11}$$ Simplify fraction to lowest terms.

Multiply.

9. $\frac{4}{5} \times \frac{15}{18}$　　**10.** $\frac{21}{16} \times \frac{26}{49}$　　**11.** $\frac{1}{3} \times 99$　　**12.** $\frac{3}{4} \times \frac{75}{80}$

To find the reciprocal of a number, write the number as a fraction. Then interchange the numerator and the denominator.

EXAMPLE 5

Find the reciprocal of $3\frac{4}{9}$.

SOLUTION

$$3\frac{4}{9} = \frac{31}{9}$$ Write $3\frac{4}{9}$ as a fraction.

$$\frac{31}{9} \Rightarrow \frac{9}{31}$$ Interchange numerator and denominator.

The reciprocal of $3\frac{4}{9}$ is $\frac{9}{31}$.

Find the reciprocal.

13. $\frac{7}{13}$　　　　**14.** $3\frac{2}{5}$　　　　**15.** $2\frac{12}{17}$　　　　**16.** $4\frac{6}{15}$

To divide by a fraction, multiply by its reciprocal.

(continued)

Algebra 1
Basic Skills Workbook: Diagnosis and Remediation

NAME _____ DATE _____

Fraction Operations

EXAMPLE 6

Divide $\frac{5}{8} \div \frac{10}{12}$.

SOLUTION

$\frac{5}{8} \div \frac{10}{12} = \frac{5}{8} \times \frac{12}{10}$ The reciprocal of $\frac{10}{12}$ is $\frac{12}{10}$.

$= \frac{5 \times 12}{8 \times 10}$ Multiply numerators and denominators.

$= \frac{60}{80}$ Simplify.

$= \frac{20 \cdot 3}{20 \cdot 4}$ Factor numerator and denominator.

$= \frac{3}{4}$ Simplify fraction to lowest terms.

Divide.

17. $\frac{7}{5} \div \frac{5}{7}$

18. $18 \div \frac{9}{10}$

19. $5\frac{1}{4} \div \frac{7}{16}$

Mixed Review

Compare the two numbers. Write the answer using <, =, or >.

20. 4995 and 4989

21. 12.698 and 12.693

22. $\frac{2}{3}$ and $\frac{24}{36}$

23. What is the area of a rectangle that is 4 ft long and 18 in. wide?

Algebra 1
Basic Skills Workbook: Diagnosis and Remediation

NAME _____ DATE _____

Quick Check

Review of Topic 1, Lesson 3

Standardized Testing Quick Check

1. To pay for a pair of pants that cost $24 and a sweater that costs $22, you hand the sales clerk a 50-dollar bill. How much change should the clerk give you?

   A. $96

   B. $4

   C. $26

   D. $2

2. Evaluate the expression $205.76 - 146.97 + 54.16$.

   A. 406.89

   B. 4.63

   C. 112.95

   D. 112.84

Homework Review Quick Check

In Exercises 3–6, add or subtract.

3. $35 + 225$

4. $48.52 - 27.47$

5. $4000 + 400 + 4$

6. $600.49 - 60.33 - 6.76$

Practice

For use before Lesson 1.4: Fraction Operations

For Exercises 1–12, find the sum or difference.

1. $\frac{3}{6} + \frac{4}{6}$

2. $\frac{2}{4} - \frac{2}{5}$

3. $\frac{6}{9} - \frac{1}{6}$

4. $\frac{1}{7} + \frac{1}{14}$

5. $\frac{5}{7} - \frac{1}{3}$

6. $\frac{1}{2} + \frac{1}{5} + \frac{3}{10}$

7. $7\frac{3}{5} + 5\frac{1}{5}$

8. $5\frac{5}{12} - 2\frac{7}{8}$

9. $8\frac{2}{3} - 1\frac{2}{9}$

10. $2\frac{7}{10} + 8\frac{1}{2}$

11. $6\frac{2}{5} - 2\frac{1}{3}$

12. $15\frac{1}{16} - 12\frac{3}{4}$

For Exercises 13–15, find the reciprocal of the number.

13. $\frac{1}{2}$

14. 25

15. $5\frac{7}{8}$

For Exercises 16–30, multiply or divide.

16. $\frac{2}{3} \times 9$

17. $4 \times \frac{6}{11}$

18. $\frac{5}{7} \times \frac{3}{4}$

19. $\frac{4}{9} \times \frac{1}{2} \times \frac{1}{3}$

20. $\frac{4}{8} \times \frac{3}{5} \times \frac{1}{2}$

21. $\frac{7}{8} \times \frac{2}{7}$

22. $4\frac{2}{7} \times \frac{1}{5}$

23. $1\frac{1}{8} \times 5\frac{1}{2}$

24. $\frac{13}{4} \div \frac{2}{3}$

25. $2\frac{7}{8} \div \frac{1}{2}$

26. $\frac{5}{8} \div \frac{1}{6}$

27. $\frac{22}{5} \div \frac{1}{3}$

28. $2\frac{2}{5} \div 2\frac{1}{3}$

29. $4\frac{1}{6} \div 5$

30. $9\frac{1}{4} \div \frac{3}{8}$

31. You need $2\frac{1}{2}$ cups of flour for your favorite cookie recipe. The only measuring cup you have holds $\frac{1}{4}$ cup of flour. How many times do you have to fill the measuring cup?

32. You want to double the cookie recipe in Exercise 31. How many cups of flour do you need?

NAME _____ DATE _____

Assessment

For use with Topic 1: Whole Numbers, Decimals, and Fractions

Find the greatest common factor and the least common multiple of the pair of numbers.

  1. 26, 27   **2.** 4, 32   **3.** 31, 7

  4. 44, 144   **5.** 60, 75   **6.** 18, 324

Compare the two numbers. Write the answer using <, =, or >.

  7. $\frac{7}{9}$ and $\frac{2}{3}$ **8.** $\frac{3}{2}$ and $1\frac{1}{2}$ **9.** $\frac{7}{11}$ and $\frac{5}{8}$

Write the numbers in order from least to greatest.

  10. $\frac{3}{2}, \frac{4}{7}, \frac{3}{4}, \frac{1}{7}$ **11.** 0.054, 0.053, 0.0035, 0.0053   **12.** 2.61, 2.4, 2.345

Find each sum.

  13. $35 + 94$ **14.** $46.3 + 52.689$ **15.** $\frac{5}{6} + 2\frac{1}{6}$

  16. $\frac{7}{18} + \frac{2}{3}$ **17.** $3\frac{5}{9} + 8\frac{1}{2}$ **18.** $\frac{31}{40} + 2\frac{7}{8}$

Find each difference.

  19. $75 - 29$ **20.** $115.98 - 57.09$ **21.** $\frac{4}{5} - \frac{2}{5}$

  22. $\frac{17}{21} - \frac{29}{42}$ **23.** $8\frac{2}{3} - 4\frac{5}{6}$ **24.** $\frac{11}{12} - \frac{9}{16}$

Find each product.

  25. 387×25 **26.** 84.26×3 **27.** $\frac{7}{3} \times \frac{6}{21}$

  28. 187.36×10.4 **29.** $\frac{4}{11} \times \frac{5}{13}$ **30.** 0.33×59.99

Find each quotient.

  31. $189 \div 5$ **32.** $68.22 \div 0.2$ **33.** $\frac{7}{16} \div \frac{1}{4}$

  34. $38 \div 1.9$ **35.** $4\frac{1}{2} \div \frac{2}{3}$ **36.** $29.3904 \div 12.56$

  37. To convert kilograms to pounds, multiply the number of kilograms by 2.2. What is the weight in pounds of a dog that weighs 15.42 kilograms?

  38. You run $5\frac{1}{2}$ times around a quarter-mile track. How far did you run?

Algebra 1
Basic Skills Workbook: Diagnosis and Remediation

Mean, Median, Mode and Range

GOAL **Find the mean, the median, the mode, and the range of a data set.**

> Caryl has a part-time summer job. Over the last five weeks, she has worked the following number of hours per week: 22, 21, 35, 21, and 26. *Statistical measures* can be used to describe the data.

Terms to Know	Example/Illustration
Mean the sum of the data in a data set divided by the number of items	$\dfrac{22 + 21 + 35 + 21 + 26}{5} = \dfrac{125}{5} = 25$ The mean number of hours per week is 25.
Median the middle number or the average of the two middle numbers in a data set when the data are arranged in numerical order	In numerical order, the data are: 21, 21, 22, 26, 35. The median number of hours per week is 22, the middle number in the data set.
Mode the most frequently occurring item, or items, in a data set (*Note:* Some data sets do not have a mode.)	The number 21 occurs more often than any other number in the data set above. Therefore, the mode of the data is 21 hours per week.
Range the difference between the greatest and the least values in a data set	The greatest value in the data set is 35 and the least value is 21. Therefore, the range of the data set is 35 − 21, or 14 hours per week.

Understanding the Main Ideas

To find the mean of a set of data, first find the sum of the data items. Then divide this sum by the number of data items in the set.

EXAMPLE 1
_____

Six single-family homes were sold in one day in the city of Westford. Find the mean selling price of the homes.

Location	Selling price ($)
Carlisle Road	162,000
Stanton Street	129,900
Pond Street	125,000
King Circle	292,400
Briar Hill Road	178,250
Washington Street	237,000

(continued)

Algebra 1
Basic Skills Workbook: Diagnosis and Remediation

Mean, Median, Mode, and Range

SOLUTION

Step 1: Find the sum of the data items.

$$
\begin{array}{r}
162,000 \\
129,900 \\
125,000 \\
292,400 \\
178,250 \\
+\ 237,000 \\
\hline
1,124,550
\end{array}
$$

Step 2: Divide the sum from Step 1 by the number of items. There are six data items, so divide by 6.

$$\frac{1,124,550}{6} = 187,425$$

The mean selling price of these homes was $187,425.

Find the mean of each data set.

1. 0.8, 1.2, 0.75, 0.62, 0.88

2. 108, 97, 86, 99, 102, 90

3. 98.8, 99.5, 102.1, 101.5, 99.1

4. 4.50, 6.25, 5.25, 10, 8.75, 8.75

To find the median of a set of data, first write the data in numerical order (either least to greatest or greatest to least). The median is the middle item if there is an odd number of data items. If there is an even number of items, then the median is the average of the two middle data items.

EXAMPLE 2

Find the median selling price for the homes in Example 1.

SOLUTION

Step 1: Write the data items in numerical order.

125,000   129,900   162,000   178,250   237,000   292,400

Step 2: Since there are six data items, identify the two middle items. Then find the average of these two items.

The two middle data items are 162,000 and 178,250.

$$\frac{162,000 + 178,250}{2} = \frac{340,250}{2} = 170,125$$

The median selling price of these homes was $170,125.

(continued)

Topic 2

NAME _____ DATE _____

Mean, Median, Mode, and Range

Find the median of each data set.

5. 142, 130, 223, 178, 129

6. 12.5, 13.4, 17.7, 20.2, 15.4, 16.1

7. 19, 17, 24, 23, 28, 35, 38

8. 72, 74, 68, 65, 70, 70, 68

To find the mode of a set of data, identify the item or items that occur most frequently in the data set. Not only can there be more than one mode, a data set can also have no mode.

EXAMPLE 3

Find the mode or modes of the selling price for the homes in Example 1.

SOLUTION

No data item occurs more than once. Therefore, there is no mode for this data set.

Find the mode or modes of each data set.

9. 0.5, 0.33, 0.7, 0.61, 0.4

10. 40, 35, 35, 38, 37, 42, 37

11. 3, 2, 4, 4, 3, 1, 2, 1, 5, 4, 6

12. 35, 54, 17, 18, 42, 44, 42, 61

To find the range of a set of data, identify the greatest and least values in the data set and subtract them.

EXAMPLE 4

Find the range of the selling prices for the homes in Example 1.

SOLUTION

The greatest selling price is $292,400 and the least is $125,000.

$$292,400 - 125,000 = 167,400$$

The range of the selling prices of these homes is $167,400.

Find the range of each data set.

13. 88, 100, 122, 91, 89, 77

14. $3\frac{1}{2}, 5\frac{3}{8}, 1\frac{1}{4}, 4\frac{3}{5}, 7\frac{1}{2}$

15. 1105, 880, 1320, 1500, 800

16. 7.2, 1.8, 1.05, 6.43, 9.2

Mixed Review

17. $751.8 - 190.73$

18. $3\frac{3}{4} + 8\frac{1}{2} + 11\frac{5}{8} + 5$

Topic 2

NAME _____ DATE _____

Quick Check

Review of Topic 1, Lesson 4

Standardized Testing Quick Check

1. Jane and Kate volunteer to stuff envelopes for a local non-profit organization. Jane stuffs $\frac{2}{5}$ of the envelopes and Kate stuffs $\frac{1}{3}$ of the envelopes. What fraction of the envelopes still need to be stuffed?

   A. $\frac{11}{15}$ **B.** $\frac{4}{15}$ **C.** $\frac{1}{15}$ **D.** $1\frac{4}{15}$

2. Which of the following does *not* have a sum of 1?

   A. $\frac{5}{6} + \frac{1}{9} + \frac{1}{18}$ **B.** $\frac{1}{2} + \frac{1}{3} + \frac{1}{6}$ **C.** $\frac{4}{16} + \frac{3}{4}$ **D.** $\frac{1}{8} + \frac{1}{4} + \frac{1}{16}$

Homework Review Quick Check

3. Divide $\frac{9}{14}$ by $\frac{12}{17}$.

4. Multiply $\frac{4}{5} \times \frac{1}{2} \times \frac{7}{8}$.

NAME _____ DATE _____

Practice

For use with Lesson 2.1: Mean, Median, Mode, and Range

Find the mean, the median, the mode(s), and the range for each set of data. If necessary, round your answers to the nearest tenth.

1. Wins by teams in the National Basketball Association's Western Conference prior to the All Star break:

   33, 32, 23, 16, 16, 11, 9, 35, 32, 25, 21, 19, 17, 17

2. Mortgage rates (percentages) at selected banks:

   7.07, 7.5, 7.37, 7.53, 7.42, 7.66, 8.13, 7.49

3. Ages of employees at The Home Run Diner:

   54, 24, 27, 18, 19, 36, 18

4. Prices (in dollars) of Web browser software programs:

   0, 0, 49.95, 24.95, 38.97, 29.95

5. Calories per cup of different types of milk products:

   160, 120, 100, 88, 88, 100

6. High temperatures (in °F) in Fairview last week:

   64, 70, 80, 72, 68, 82, 70

Find the mean, the median, the mode(s), and the range for the numerical data in each table. If necessary, round your answers to the nearest tenth.

7. **Sodium Content of Some Vegetables**

Vegetable	Sodium (mg/serving)
artichokes	46
beets	81
carrots	51
celery	151
kale	47
spinach (cooked)	90

8. **Population in the Five Most Densely Populated States (1990 Census)**

State	Persons per square mile
Connecticut	678.4
Maryland	489.2
Massachusetts	767.6
New Jersey	1042.0
Rhode Island	960.3

Topic 2

NAME _____ DATE _____

Bar Graphs and Line Graphs

GOAL | **Read and interpret bar graphs and line graphs. Construct a bar graph or a line graph for a data set.**

Data may be displayed using a graph. The type of graph that is most effective depends on the nature of the data.

Terms to Know

Example/Illustration

Bar graph
a type of graph in which the lengths of vertical or horizontal bars are used to compare data that fall into different categories

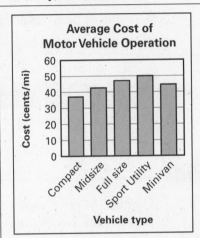

Line graph
a type of graph in which data points connected by line segments show the change in the data over time

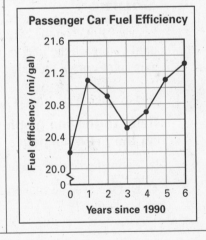

Understanding the Main Ideas

One axis (most often the vertical axis) of a bar graph is marked with a numerical scale. The other axis is labeled with several categories. The data are represented by the lengths of the bars.

EXAMPLE 1

Refer to the bar graph shown in the Terms to Know. Use the graph to estimate the average operating cost of a minivan.

(continued)

Algebra 1
Basic Skills Workbook: Diagnosis and Remediation

Topic 2

NAME _____ DATE _____

Bar Graphs and Line Graphs

SOLUTION

Locate the bar labeled "Minivan." The top of the bar is just below 45 on the vertical scale. Therefore, the graph shows that the cost of operating a minivan is about 45¢/mi.

Refer to the bar graph shown in the Terms to Know on the previous page.

1. Estimate the average cost of operating a midsize car.

2. Compare the operating costs of full-size cars and compacts.

In many line graphs, one axis is a numerical scale. The other axis usually represents a period of time. A gap in a scale is indicated by a jagged line. The data in a line graph are represented by points connected by line segments.

EXAMPLE 2

Use the line graph shown in the Terms to Know on the previous page. Estimate the average fuel efficiency for passenger cars in 1993.

SOLUTION

Locate "1993" on the horizontal axis. Move straight up from "1993" until you reach the line. From this point, move directly left to the vertical axis and estimate the value on the scale: about 20.5.

The average fuel efficiency for passenger cars in 1993 was about 20.5 mi/gal.

Refer to the line graph shown in the Terms to Know on the previous page.

3. Estimate the average fuel efficiency for passenger cars in 1991.

4. Did the average fuel efficiency increase or decrease from 1993 to 1994? by how much?

You can draw bar graphs to display data involving different categories.

EXAMPLE 3

Draw a bar graph to display the data in the table.

Annual Snowfall at Colorado Ski Areas

Ski Area	Aspen Mt.	Bear Creek	Keystone	Telluride	Vail
Snowfall (in.)	300	330	230	300	335

(continued)

Algebra 1
Basic Skills Workbook: Diagnosis and Remediation

LESSON
2.2
CONTINUED

Bar Graphs and Line Graphs

SOLUTION

Begin by writing the five ski areas along the horizontal axis. Then choose a numerical scale for the vertical axis. Since the greatest data value is 335 in., a scale from 0 in. to 350 in. with intervals of 50 in. is used here.

Now use the data in the table to draw a bar for each ski area. Finally, write a title for the graph.

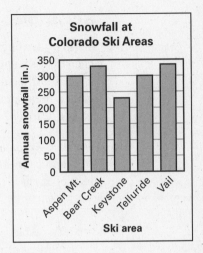

5. Only seven states are the birthplace of more than one United States president. The table below shows the number of presidents born in each of those seven states. Draw a bar graph to display the data.

State	Number of presidents
Massachusetts	4
North Carolina	2
New York	4
Ohio	7
Texas	2
Vermont	2
Virginia	8

You can draw a line graph to display data involving a single category for which the value changes over time.

EXAMPLE 4

Draw a line graph to display the data in the table below.

U.S. Patent Applications

Year	1991	1992	1993	1994	1995
Number of applications (thousands)	178	185	188	201	236

(continued)

Topic 2

Bar Graphs and Line Graphs

SOLUTION

Begin by writing the years along the horizontal axis. Then choose a numerical scale for the vertical axis. Since all of the data values are between 178 and 236, show a break in the scale after 0 and then show values from 170 to 270 using intervals of 20.

Now use the data in the table to plot a point for each year, and connect the five points from left to right with line segments. Finally, write a title for the graph.

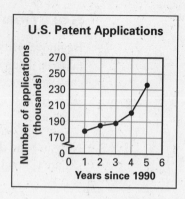

6. The table at the right shows the number of personal computers in use in the United States each year from 1991 to 1995. Draw a line graph to display the data.

7. *Writing* Write a paragraph explaining how the data in Exercise 6 can also be displayed in a bar graph.

Year	Number of personal computers in the U.S. (nearest million)
1991	59
1992	65
1993	73
1994	82
1995	92

Mixed Review

8. Find the mean, the median, the mode(s), and the range of the data in Exercise 5 on the previous page. Round your answers to the nearest tenth if necessary.

9. Find the least common multiple and greatest common factor of 35 and 28.

10. List all the common factors of 32 and 48.

Algebra 1
Basic Skills Workbook: Diagnosis and Remediation

Topic 2

NAME _____ DATE _____

Quick Check

Review of Topic 2, Lesson 1

Standardized Testing Quick Check

1. The low temperatures (°F) recorded in Erie, PA, one week were
   50°, 46°, 52°, 48°, 48°, 45°, and 42°. What is the median
   temperature?

   A. 47.29°   **B.** 7°

   C. 48°   **D.** 52°

2. In a survey of 15 people, the number of telephones in each person's
   home were as follows.

   3, 1, 2, 2, 4, 3, 3, 2, 3, 3, 4, 3, 2, 3, 3

   What is the mode?

   A. 2.73   **B.** 3   **C.** 2   **D.** 1

Homework Review Quick Check

**Find the mean, the median, the mode(s), and the
range of each data set.**

3. Vitamin C (mg) in single servings of various fruits:

   7, 12, 2, 66, 45

4. Number of languages spoken in certain high schools in 1996:

   23, 20, 31, 55, 33, 33

NAME _____ DATE _____

Practice

For use with Lesson 2.2: Bar Graphs and Line Graphs

For Exercises 1–6, use the bar graph at the right.

For Exercises 1–3, estimate the number of calories used per minute for each activity.

1. walking at a rate of 3 mi/h

2. tennis

3. jogging

4. For which of the listed activities would a 120 lb person use more than 5 calories per minute?

5. Which activity produces a calorie use about twice that of volleyball?

6. About how many calories would a 120 lb person use if he or she rode a bicycle at a rate of 5 mi/h for half an hour?

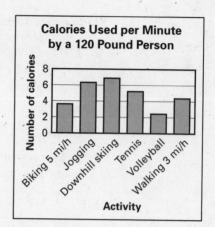

For Exercises 7–11, use the line graph.

For Exercises 7 and 8, estimate the number of computers per 1000 people each year.

7. 1992

8. 1994

9. In what year were there about 35 computers per 1000 people world-wide?

10. In what year did the number of computers per 1000 people pass 40 for the first time?

11. By approximately how much did the number of computers per 1000 people increase from 1991 to 1995?

12. Draw a bar graph to display the data in the table.

Sites of the Olympic Games by Continent (1896–2002)

Continent	Asia	Australia	Europe	North America
Games held	4	2	27	11

13. Draw a line graph to display the data in the table.

Outstanding Consumer Credit in the United States

Year	1990	1991	1992	1993	1994	1995
Amount owed (billions of dollars)	752	745	757	807	925	1132

NAME _____ DATE _____

Circle Graphs

Topic 2

GOAL   **Read and interpret circle graphs.**

A circle graph is a very useful display when you want to show how each part of something relates to the whole thing, as well as to show how the parts relate to each other.

Terms to Know	**Example/Illustration**

Circle graph
    a type of graph in which the data are
    represented as parts of a whole
	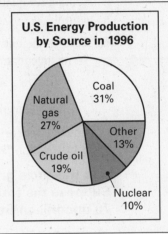
	U.S. Energy Production by Source in 1996

Sector
    a wedge-shaped part of a circle; in a
    circle graph, a region that represents
    part of the data being displayed
	The circle graph above contains five sectors.

Understanding the Main Ideas

The data in a circle graph are expressed as parts (or percentages) of a whole. The entire circle represents 100% of the data and each sector represents a certain percentage of the data.

EXAMPLE 1

Use the circle graph shown in the Terms to Know.

A. If the total amount of energy produced in the United States in 1996 was about 70 quadrillion British thermal units (Btu's), about how much energy was produced by burning natural gas?

B. Suppose nuclear energy production accounted for 7.1 quadrillion Btu's in 1996. What was the total number of Btu's of energy produced in the United States in 1996?

(continued)

Circle Graphs

SOLUTION

A. The circle graph shows that the amount of energy produced in the United States in 1996 by burning natural gas was 27% of the total energy produced.

$$70 \times 0.27 = 18.9$$

About 18.9 quadrillion Btu's were produced in the United States by burning natural gas.

B. Let t = the total number of Btu's produced in the United States in 1996. Write an equation.

10% of $t = 7.1 \rightarrow 0.1 \times t = 7.1$

Divide to find the missing factor: $t = 7.1 \div 0.1 = 71$

If nuclear energy accounted for 7.1 quadrillion Btu's, then a total of about 71 quadrillion Btu's of energy was produced in the United States in 1996.

Use the circle graph in the Terms to Know on the previous page. Suppose the total U.S. energy production in 1996 was about 70 quadrillion Btu's.

1. Estimate the amount of energy produced in the United States by burning crude oil.

2. About how much energy was produced by burning fossil fuels, that is, coal, natural gas, and crude oil?

Circle graphs can also be drawn showing the actual data values in the sectors, rather than percents.

EXAMPLE 2

Use the circle graph at the right. The graph shows the amount of energy of various types (in quadrillion Btu's) that was consumed in the United States in 1996.

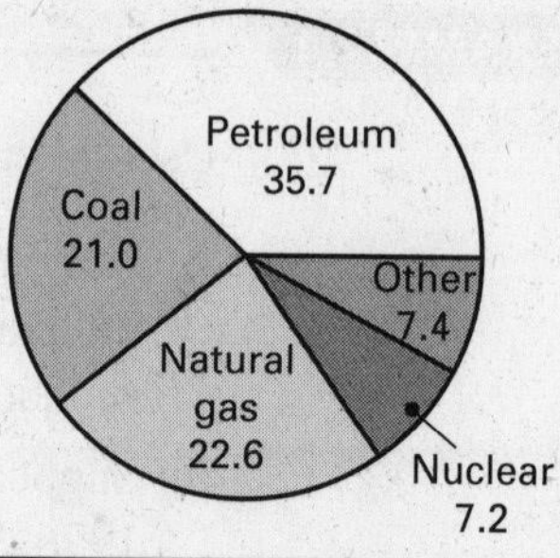

A. What was the total amount of energy consumed in the United States in 1996?

B. What percent of the energy consumed in 1996 was accounted for by natural gas?

Energy Consumption by Source in 1996 in Quadrillion Btu's

Petroleum 35.7
Coal 21.0
Other 7.4
Natural gas 22.6
Nuclear 7.2

SOLUTION

A. Add the amounts shown in the five sectors.

$$35.7 + 21.0 + 22.6 + 7.2 + 7.4 = 93.9$$

The total amount of energy consumed in the United States in 1996 was 93.9 quadrillion Btu's.

(continued)

Topic 2

NAME _____ DATE _____

Circle Graphs

B. The total amount consumed was 93.9 quadrillion Btu's, while the total natural gas consumption was 22.6 quadrillion Btu's.

$$22.6 \div 93.9 \approx 0.241$$

So, natural gas accounted for about 24.1% of the energy consumed in the United States in 1996.

Use the circle graph in Example 2. Find the percent of the total energy consumption in the United States in 1996 for each energy source. Round your answers to the nearest tenth of a percent.

3. nuclear

4. other

5. all fossil fuels (petroleum, natural gas, coal)

Mixed Review

Evaluate the expression.

6. $\dfrac{78 + 88 + 72 + 89 + 88}{5}$

7. $\dfrac{86 + 112 + 35 + 18 + 9}{5}$

8. $\dfrac{2}{5} - \dfrac{3}{10}$

9. $68 \times \dfrac{12}{17}$

10. $117.03 - 59.788$

11. $56.875 \div 16.25$

Quick Check

Review of Topic 2, Lesson 2

Standardized Testing Quick Check

1. A nurse records a patient's temperature every hour for 24 h. Which type of graph would best display the data?

   A. bar graph

   B. circle graph

   C. line graph

   D. scatter plot

Homework Review Quick Check

Use the bar graph at the right.

2. About how many mammal species are endangered?

3. Estimate the difference between the number of endangered bird species and the number of endangered reptile species.

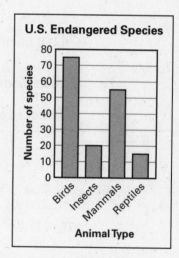

Use the line graph at the right.

4. Estimate the population of the United States in 1970.

5. By about how much did the population of the United States increase between 1950 and 1990?

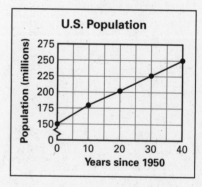

Algebra 1
Basic Skills Workbook: Diagnosis and Remediation

NAME _____ DATE _____

Practice

For use with Lesson 2.3: Circle Graphs

Private health insurance falls into the four categories shown in the circle graph at the right. For Exercises 1–4, estimate the number of persons with each type of insurance among a group of 100,000 privately insured people.

1. private care

2. Health Maintenance Organization (HMO)

3. managed care—choice of doctor

4. managed care—specified doctor

5. Use the circle graph above. If about 15,000 privately insured persons in a large group have managed care in which they may choose their own doctors, estimate the size of the entire group.

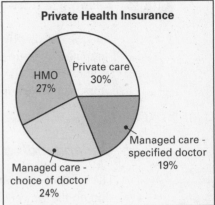

Private Health Insurance

HMO 27%
Private care 30%
Managed care - specified doctor 19%
Managed care - choice of doctor 24%

The number of motor vehicles registered in the United States in 1995 was about 206,000,000. For Exercises 6–8, use the circle graph at the right to estimate the number of each type of vehicle.

6. automobiles

7. trucks and buses (combined)

8. motorcycles

9. If a circle graph was drawn just for the number of motor vehicles registered in the state of Michigan, the percentages would be the same as those in the circle graph above. In 1995, there were 5,045,000 automobiles registered in Michigan. Estimate the total number of registered motor vehicles in Michigan to the nearest thousand.

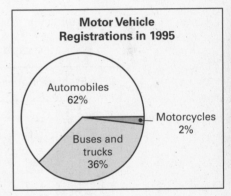

Motor Vehicle Registrations in 1995

Automobiles 62%
Motorcycles 2%
Buses and trucks 36%

Use the circle graph at the right.

10. Find the number of ski runs at Alpine Cliffs.

11. What percent of the ski runs at Alpine Cliffs is in each category?

    a. expert

    b. intermediate

    c. beginner

12. *Open-ended*   Why do you think that almost half of the ski runs are for intermediate-level skiers?

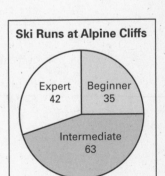

Ski Runs at Alpine Cliffs

Expert 42
Beginner 35
Intermediate 63

Topic 2

NAME _____ DATE _____

Interpreting Graphs

GOAL
Recognize how graphs can be misleading.

In Lesson 2.2, you learned how to display data using bar graphs and line graphs. Depending on how they are drawn, two bar graphs or two line graphs that display exactly the same data can create two different visual impressions. Some graphs may even be misleading to the person viewing them.

Understanding the Main Ideas

The scale of a graph can be chosen in such a way as to make a desired impression on the viewer.

EXAMPLE 1 _____

Both of these bar graphs display the same data.

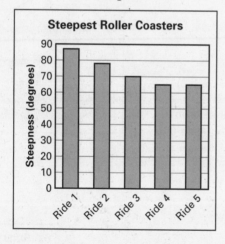
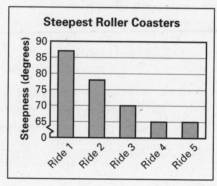

A. Explain why Graph B could be misleading.

B. Explain why an advertisement about Ride 1 would more likely include Graph B than Graph A.

(continued)

NAME _____ DATE _____

Interpreting Graphs

SOLUTION

A. Because of the scale on Graph B, the relative lengths of the bars are misleading. For example, the bar for the steepest roller coaster in the group, Ride 1, is five times as long as those for the two least steep roller coasters. The graph implies that Ride 1 is five times as steep as Ride 4 and Ride 5, but it is actually much less than twice as steep.

B. Graph B overemphasizes the steepness of Ride 1, making it seem more thrilling.

Use the graphs in Example 1.

1. Describe the impression that Graph B gives of the steepness of Ride 3 and Ride 5.

2. Explain why the owners of Ride 3 might use Graph A in an advertisement.

A line graph may be misleading if it makes the change in the data over time appear to be more dramatic than it actually is.

EXAMPLE 2

Both line graphs below show the median income for a family of four in the United States from 1988 to 1994.

Graph C

Graph D

A. Compare the visual impressions made by the two graphs.

B. Which graph would you use to support an argument that the economic situation in the United States improved greatly during the period from 1988 to 1994?

(continued)

Algebra 1
Basic Skills Workbook: Diagnosis and Remediation

NAME _____ DATE _____

Interpreting Graphs

Topic 2

SOLUTION

A. Graph C makes it appear that the median income for a family of four has improved slowly but steadily from 1988 to 1994. Graph D makes it appear that the median income for a family of four increased sharply over that same period of time.

B. Graph D; this graph makes the increases from one year to the next seem more dramatic than they appear in Graph C.

Use the graphs in Example 2.

3. Under what circumstances do you think a politician running for reelection would use Graph D in his or her campaign literature?

4. Look at the labels on the horizontal axis of Graph C. What effect do you think spacing the labels farther apart would have on the impression given by the line graph?

Mixed Review

In Exercises 5–7, use the bar graph at the right.

5. How many gold medals were won by Germany?

6. Which country won about half as many gold medals as the United States?

7. How many countries won more than 20 gold medals?

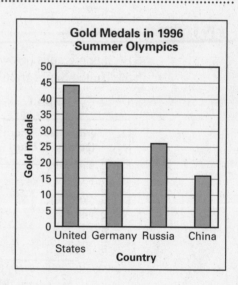

NAME _____ DATE _____

Quick Check

Review of Topic 2, Lesson 3

Standardized Testing Quick Check

1. The kinds of books purchased by 16,000 adults in 1996 are shown in the circle graph at the right. Use the graph to estimate how many more people bought popular fiction books than bought general nonfiction books.

   A. 6720   **B.** 6560

   C. 3360   **D.** 6240

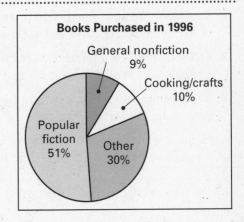

Books Purchased in 1996

General nonfiction 9%
Cooking/crafts 10%
Popular fiction 51%
Other 30%

Homework Review Quick Check

Use the circle graph at the right.

2. What percent of the forest land had forest on it?

3. There were 191 million acres of National Forest land in 1995. Estimate the number of acres on which timber harvest was permitted.

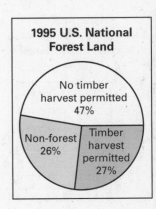

1995 U.S. National Forest Land

No timber harvest permitted 47%
Non-forest 26%
Timber harvest permitted 27%

Topic 2

NAME _____ DATE _____

Practice

For use with Lesson 2.4: Interpreting Graphs

Use the bar graph at the right.

1. Describe the visual impression of the relationship between gourmet coffee prices at Alonzo's and at Cafe Break.

2. Describe the visual impression of the relationship among gourmet coffee prices at all four stores.

3. Which store do you think is most likely to use this graph in its advertising? Which is least likely to use it?

4. Explain how the graph could be redrawn to give a more favorable impression of coffee prices at Cafe Break.

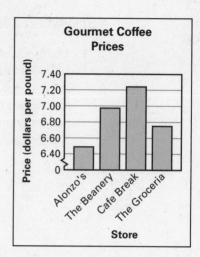

Use these line graphs for Exercises 5 and 6.

Graph A   **Graph B**

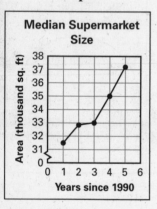   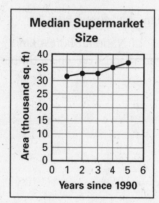

5. Compare the visual impressions given by the two graphs.

6. A developer is trying to convince investors to increase the size of a planned supermarket. Which graph should the developer use? Explain.

Use the bar graph at the right. Tell whether each statement is *True* or *False*.

7. The game *Star Traveler* costs almost twice as much at Comp-Town as it does at Romulus.

8. The game costs only half as much at CD Land as it does at Game Shack.

9. The difference between the highest and lowest prices for the game is $7.

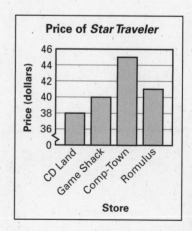

Algebra 1
Basic Skills Workbook: Diagnosis and Remediation

NAME _____ DATE _____

Assessment

For use with Topic 2: Working with Data

Find the mean, the median, and mode(s), and the range of each data set.

1. Runs scored in six games of the World Series:

   13, 4, 7, 14, 1, 5

2. Homeroom attendance for one week:

   30, 28, 32, 31, 30

Use the bar graph at the right.

3. Estimate the amount of money Graf won in 1996.

4. Which player's winnings were equal to about half of Graf's?

5. How many women tennis players won more than $1 million in 1996?

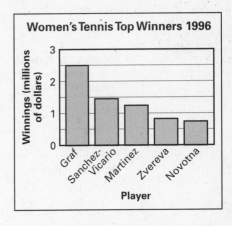

For Exercises 6 and 7, use the line graph at the right.

6. Estimate the life expectancy for a male born in the United States in 1980.

7. Over what ten-year period shown in the graph did the life expectancy for males born in the United States increase the most?

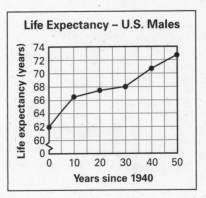

8. Determine whether the data in the table below should be displayed in a bar graph or a line graph. Then graph the data.

Pull-ups Required for Physical Fitness Award (Boys)

Age	12	13	14	15	16	17
Number of pull-ups	7	7	10	11	11	13

9. Use the circle graph at the right. Suppose charitable donations in 1996 totaled $150 billion. How much of that was donated by corporations?

10. Explain why you can compare the data in Exercises 3–5 above by comparing the lengths of the bars.

11. Explain how you could alter the scale on the graph for Exercises 3–5 to make Graf's winnings even more impressive.

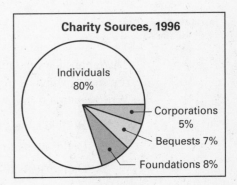

Rates and Ratios

GOAL **Find the unit rates from words and graphs. Use ratios and rates to compare quantities.**

> In 1990, about 51 of every 100 people in the United States were females. Therefore, about 49 of every 100 people were males. The ratio of females to males was about $\frac{51}{49}$.

Terms to Know	Example/Illustration
Ratio of a to b the relationship $\frac{a}{b}$ of two quantities a and b that are measured in the same units	51 to 49, $\frac{51}{49}$, 51:49
Rate of a per b the relationship $\frac{a}{b}$ of two quantities a and b that are measured in different units	100 miles per 2 hours, $\dfrac{100 \text{ miles}}{2 \text{ hours}}$
Unit Rate a rate per one given unit	50 miles per hour, $\dfrac{50 \text{ miles}}{1 \text{ hour}}$

Understanding the Main Ideas

The ratio of one number a to another number b $(b \neq 0)$ is the quotient when a is divided by b. There are three ways to express the ratio:

$$a \text{ to } b \qquad\qquad \frac{a}{b} \qquad\qquad a:b$$

To write a ratio in lowest terms, simply write the ratio as a fraction in lowest terms. However, do not change the fraction to a mixed number. If the simplified fraction has a denominator of 1, be sure to write it as the second value in the ratio.

EXAMPLE 1

The table shows the records of four of the 1995 division champions in the National Football League. Write each ratio in lowest terms.

   A. Buffalo's wins to losses

   B. Dallas' wins to losses

   C. Dallas' wins to Buffalo's wins

Team	Wins	Losses
Buffalo	10	6
Kansas City	13	3
Dallas	12	4
San Francisco	11	5

(continued)

Rates and Ratios

SOLUTION

A. $\frac{10}{6} = \frac{5}{3}$ **B.** $\frac{12}{4} = \frac{3}{1}$ **C.** $\frac{12}{10} = \frac{6}{5}$

Refer to the table in Example 1. Write each ratio in lowest terms.

1. Kansas City's wins to losses

2. Buffalo's losses to Dallas's losses

3. Buffalo's losses to Kansas City's losses

If a ratio compares two like quantities measured in different units, first write the measurements so the units are the same.

EXAMPLE 2

Write the ratio of 27 yd to 18 ft in lowest terms.

SOLUTION

Rewrite the measures so they have the same units.
There are two choices: (1) rewrite in feet or (2) rewrite in yards.

(1) Change 27 yd to feet: since 1 yd = 3 ft, 27 yd = 27 × 3 ft, or 81 ft.

$$\frac{81 \text{ ft}}{18 \text{ ft}} = \frac{81}{18} = \frac{9}{2}$$

(2) Change 18 ft to yards: since 1 ft = $\frac{1}{3}$ yd, 18 ft = 18 × $\frac{1}{3}$ yd, or 6 yd.

$$\frac{27 \text{ yd}}{6 \text{ yd}} = \frac{27}{6} = \frac{9}{2}$$

Notice that both choices result in the same ratio.

Write each ratio as a fraction in lowest terms.

4. 5 km to 300 m   **5.** 3 weeks : 8 days   **6.** 12 ft to 8 in.

Topic 3

(continued)

NAME _____ DATE _____

Rates and Ratios

A *unit rate* compares a quantity to a single unit of a different quantity. For example, the speed of a car might be 55 miles per hour. This is the unit rate $\frac{55 \text{ mi}}{1 \text{ h}}$.

Notice that the bottom quantity is 1, and that the units (miles and hours) are different.

EXAMPLE 3

Write a unit rate for each situation.

A. $48 in 3 h

B. 180 mi in 3 h

SOLUTION

A. Use the rate $\frac{\text{dollars}}{\text{hours}}$: $\frac{48}{3} = \frac{16}{1}$.

The unit rate is $\frac{\$16}{1 \text{ h}}$, or $16 per hour ($16/h).

B. Use the rate $\frac{\text{miles}}{\text{hours}}$: $\frac{180}{3} = \frac{60}{1}$

The unit rate is $\frac{60 \text{ mi}}{1 \text{ h}}$, or 60 miles per hour (60 mi/h).

Write a unit rate for each situation.

7. $10 for 2 lb

8. 1200 mi in 4 days

9. $1160 in 4 weeks

10. 52 m in 4 s

Mixed Review

11. Explain why the line graph at the right is misleading.

12. Find the greatest common factor and least common multiple of 72 and 88.

13. You and two friends go out for pizza. The total bill comes to $16.41. If you are splitting the bill evenly, how much does each person owe?

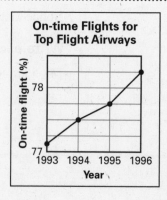

On-time Flights for Top Flight Airways

Topic 3

NAME _____ DATE _____

Quick Check

Review of Topic 2, Lesson 4

Standardized Testing Quick Check

1. Which of the following statements about the graph at the right is false?

   A. Vertical scale uses only even numbers.

   B. Vertical scale is broken.

   C. Horizontal scale is uneven.

   D. Meaning of vertical scale is unknown.

Years since 1995

Homework Review Quick Check

Use the bar graph at the right. Tell whether each statement is *True* or *False*.

2. About five times as much chocolate is consumed per person in Switzerland as in the United States.

3. The difference between the country with the greatest rate of consumption and the country with the lowest rate of consumption is 11 pounds per person.

4. Half as much chocolate is consumed per person in Belgium as in the United Kingdom.

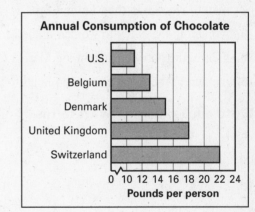

Annual Consumption of Chocolate

Pounds per person

NAME _____ DATE _____

Practice
For use with Lesson 3.1: Rates and Ratios

The table below shows the approximate number of men and women on active duty in the United States armed forces in 1995.

Branch	Number of women (thousands)	Number of men (thousands)
Army	68	507
Air Force	64	400
Navy	56	438
Marine Corps	8	174

Use the table to write each ratio in lowest terms.

1. women to men in the Air Force

2. men to women in the Marine Corps

3. women in the Navy to women in the Marine Corps

4. men in the Navy to men in the Marine Corps

5. women in the Army to women in all four branches of the armed services

Write each ratio in lowest terms.

6. 63 to 35   **7.** 120 to 80

8. 28 ft:6 in.   **9.** 6 lb to 4 oz

10. 10 m to 2 cm   **11.** 3 days to 8 h

12. 3 h:15 min   **13.** $12 to $2.50

14. 2 days to 3 h   **15.** 1 yd to 18 in.

Write the unit rate.

16. 270 mi in 6 h   **17.** $51 in 6 h

18. 16 lb in 8 weeks   **19.** $30 for 4 tickets

20. $8.25 for 3 lb   **21.** 300 words in 5 min

22. 15 m in 2 s   **23.** $1800 in 12 months

24. $21 for 15 gal   **25.** 180 km in 2 h

26. 135 revolutions in 3 min   **27.** 200 students to 8 teachers

Topic 3

NAME _____ DATE _____

Equal Rates

GOAL How to decide if two fractions, ratios, or rates are equivalent.

> In this lesson, you will learn that fractions can be written in various forms. Being able to recognize the equivalent forms of fractions, ratios, and rates will help you become a good problem solver.

Terms to Know	**Example/Illustration**
Equivalent Fraction fractions that represent the same number	$\dfrac{1}{2} = \dfrac{2}{4}$

Understanding the Main Ideas

Consider the models shown below. Notice how all the models have the same ratio of shaded parts to total parts. For instance, the first model has $\frac{1}{4}$ of the blocks shaded while the second model has $\frac{2}{8}$, or $\frac{1}{4}$ of its blocks shaded. Study the model to see that you can find equivalent fractions by multiplying the numerator and denominator by the same number.

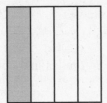 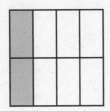 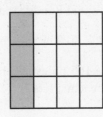 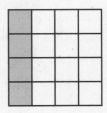

$\dfrac{1}{4} \, or$ $\dfrac{2}{8} \, or$ $\dfrac{3}{12} \, or$ $\dfrac{4}{16} \, or$

$\dfrac{1 \cdot 1}{4 \cdot 1} = \dfrac{1}{4}$ $\dfrac{1 \cdot 2}{4 \cdot 2} = \dfrac{2}{8}$ $\dfrac{1 \cdot 3}{4 \cdot 3} = \dfrac{3}{12}$ $\dfrac{1 \cdot 4}{4 \cdot 4} = \dfrac{4}{16}$

EXAMPLE 1

Show that $\dfrac{2}{5}$ and $\dfrac{6}{15}$ are equivalent fractions.

SOLUTION

$\dfrac{2}{5} = \dfrac{2 \cdot 3}{5 \cdot 3} = \dfrac{6}{15}$ **Multiply numerator and denominator by 3.**

The fractions $\dfrac{2}{5}$ and $\dfrac{6}{15}$ are equivalent.

(continued)

Algebra 1
Basic Skills Workbook: Diagnosis and Remediation

Topic 3

NAME _____ DATE _____

Equal Rates

Decide whether the fractions are equivalent. Explain why or why not.

1. $\dfrac{1}{3}, \dfrac{4}{12}$ 2. $\dfrac{2}{7}, \dfrac{16}{49}$ 3. $\dfrac{5}{6}, \dfrac{25}{30}$ 4. $\dfrac{21}{24}, \dfrac{7}{8}$

Ratios and rates are equivalent if they are equivalent as fractions.

EXAMPLE 2 _____

Mary types 3 pages in 12 minutes. Randy types 5 pages in 20 minutes. Do they type at the same rate?

SOLUTION

Mary's and Randy's rates are described in the same units, so you can see if their rates are equivalent as fractions.

$$\dfrac{3 \text{ pages}}{12 \text{ minutes}} = \dfrac{3}{12} = \dfrac{1}{4} \qquad \textbf{Mary's rate}$$

$$\dfrac{5 \text{ pages}}{20 \text{ minutes}} = \dfrac{5}{20} = \dfrac{1}{4} \qquad \textbf{Randy's rate}$$

Mary's and Randy's rate can be reduced to $\dfrac{1 \text{ page}}{4 \text{ minutes}}$.

So, they type at the same rate.

5. You drove 165 miles in 3 hours. Your friend drove 240 miles in 4 hours. Did you drive at the same rate?

You can also use equivalent fractions to find the missing number in a ratio. For example, if you know the ratio of the width to the length of the rectangle shown is 2:3 and the length is 12 inches, you can find the width as follows.

$$\dfrac{2}{3} = \dfrac{?}{12} \implies \dfrac{2 \cdot 4}{3 \cdot 4} = \dfrac{8}{12} \implies \text{The width is 8 inches.}$$

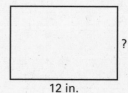

12 in.

Mixed Review

6. Describe the pattern. Then use the pattern to write the next three numbers.
   3, 6, 9, 12, 15, . . .

7. Gregory tries to balance his caloric intake from fat, protein, and carbohydrates as shown in the graph. His total intake is 2000 calories per day. How many calories does he get from protein?

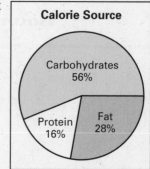

Calorie Source

Carbohydrates 56%

Protein 16%

Fat 28%

Algebra 1
Basic Skills Workbook: Diagnosis and Remediation

Topic 3

NAME _____ DATE _____

Quick Check

Review of Topic 3, Lesson 1

Standardized Testing Quick Check

1. Which is *not* a rate?

   A. 5 meters per second

   B. 3 seconds per second

   C. 10 dollars per hour

   D. 7 dollars per 2 pounds

2. You are a sales representative for a sporting goods company. You make 30 sales calls and sell $2400 of merchandise. Estimate the amount of sales, in dollars per sales call.

   A. $8 per sales call

   B. $80 per sales call

   C. $125 per sales call

   D. $800 per sales call

Homework Review Quick Check

Write each ratio in lowest terms.

3. 165 apples : 135 apples

4. 10 gallons to 15 gallons

5. 40 hours to 35 hours

6. *Movie Rentals* For the 5-year period from 1990 through 1995, Americans averaged 243 hours watching movies on video. Find the average number of hours watched per year. The average length of a movie is 1.5 hours. Estimate the number of movies watched per year.

NAME _____ DATE _____

Practice

For use with Lesson 3.2: Equal Rates

Match the fraction with its diagram. Write an equivalent fraction that is represented by the diagram.

A. 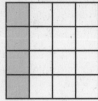 B. 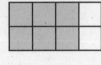 C. 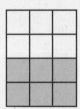 D.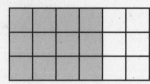

1. $\frac{2}{3}$ **2.** $\frac{3}{4}$ **3.** $\frac{1}{4}$ **4.** $\frac{1}{2}$

Find the missing number.

5. $\frac{3}{7} = \frac{?}{42}$ **6.** $\frac{5}{8} = \frac{25}{?}$ **7.** $\frac{12}{34} = \frac{?}{17}$ **8.** $\frac{2}{?} = \frac{8}{36}$

Write three equivalent fractions. Tell why they are equivalent.

9. $\frac{6}{11}$ **10.** $\frac{4}{9}$ **11.** $\frac{7}{14}$ **12.** $\frac{5}{16}$

Is the statement true or false? If it is false, change the bold number to make the statement true.

13. $\frac{7 \text{ inches}}{12 \text{ inches}} = \frac{21 \text{ inches}}{\textbf{36} \text{ inches}}$ **14.** $\frac{300 \text{ miles}}{5 \text{ hours}} = \frac{\textbf{55} \text{ miles}}{1 \text{ hour}}$

15. $\frac{18}{24} = \frac{3}{\textbf{8}}$ **16.** $\frac{4 \text{ boys}}{6 \text{ girls}} = \frac{\textbf{14} \text{ boys}}{21 \text{ girls}}$

Find the missing side of the rectangle. The ratio of the width to the length is given.

17. 2 to 5

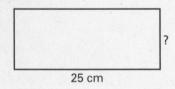

?

25 cm

18. 8 to 12

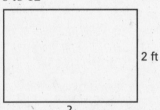

2 ft

?

19. *Running* Tyler ran 10 miles in 75 minutes. Brian ran 4 miles in 30 minutes. Did they run at the same rate? Explain.

20. *Scale Drawing* On a blueprint for a house, the ratio of the drawing to the house is 2 centimeters to 7 feet. If the wall in one bedroom has a length of 14 feet, what is the length of the wall on the blueprint?

Topic 3

54 **Algebra 1**
Basic Skills Workbook: Diagnosis and Remediation

NAME _____ DATE _____

Fractions, Decimals, and Percents

GOAL Write a fraction as a decimal or a percent, or vice versa. Write fractions in lowest terms.

> In 1990, about 51 of every 100 people in the United States were females. The number of females per 100 people in the United States can be written as a fraction $\left(\frac{51}{100}\right)$, a decimal (0.51), or a percent (51%).

Terms to Know	*Example/Illustration*
Repeating decimal a decimal in which a single digit or a block of digits repeats without end	0.234234234. . . The three dots (called an *ellipsis*) indicate that the pattern of digits continues to repeat. (*Note:* The decimal number can also be written using a bar over the repeating digits: $0.\overline{234}$.)
Percent a ratio comparing a number to 100 (Percent means "per hundred.")	$51\% = \dfrac{51}{100}$ The symbol % is read "percent."
Lowest terms the form of a fraction for which the greatest common factor (GCF) of the numerator and denominator is 1	The fraction $\frac{5}{6}$ is in lowest terms since the greatest common factor of 5 and 6 is 1.

Understanding the Main Ideas

To write a fraction as a decimal, divide the numerator by the denominator. If the remainder is not zero and a digit or a block of digits repeats, the decimal is called a *repeating decimal*.

EXAMPLE 1

Write each fraction or mixed number as a decimal.

    A. $\dfrac{7}{8}$ **B.** $2\dfrac{5}{11}$

(continued)

Algebra 1
Basic Skills Workbook: Diagnosis and Remediation

Fractions, Decimals, and Percents

SOLUTION

A.

$$8 \overline{)7.000}$$

$$\begin{array}{r} 0.875 \\ 8\overline{)7.000} \\ \underline{64} \\ 60 \\ \underline{56} \\ 40 \\ \underline{40} \\ 0 \end{array}$$

So, $\frac{7}{8} = 0.875$.

B.

$$\begin{array}{r} 0.4545\ldots \\ 11\overline{)5.000000} \\ \underline{44} \\ 60 \\ \underline{55} \\ 50 \\ \underline{44} \\ 60 \\ \underline{55} \\ 50 \end{array}$$

So, $2\frac{5}{11} = 2.\overline{45}$.

Write each fraction or mixed number as a decimal.

1. $\frac{5}{12}$ **2.** $1\frac{7}{9}$ **3.** $3\frac{4}{5}$

To write a fraction in lowest terms, divide both the numerator and denominator by their greatest common factor (GCF).

EXAMPLE 2 ————————————————————————————

Write the fraction $\frac{18}{27}$ in lowest terms.

SOLUTION

The GCF of 18 and 27 is 9.

$$\frac{18}{27} = \frac{18 \div 9}{27 \div 9} = \frac{2}{3}$$

In lowest terms, the fraction $\frac{18}{27}$ is $\frac{2}{3}$.

Write each fraction or mixed number in lowest terms.

4. $\frac{15}{60}$ **5.** $\frac{21}{35}$ **6.** $3\frac{8}{12}$

Topic 3

(continued)

Algebra 1
Basic Skills Workbook: Diagnosis and Remediation

NAME _____ DATE _____

Fractions, Decimals, and Percents

To write a fraction as a percent, rewrite it as a fraction with denominator 100, if possible. If the denominator is not a factor of 100, divide the numerator by the denominator. Then change the decimal to a percent by moving the decimal point two places to the right and attaching the percent symbol.

EXAMPLE 3
_____

Write each fraction as a percent.

A. $\dfrac{13}{25}$ 　　　　　　　　　　　**B.** $\dfrac{5}{8}$

SOLUTION

A. Since 25 is a factor of 100, rewrite the fraction with a denominator of 100. Then give the percent.

$$\frac{13}{25} = \frac{13 \cdot 4}{25 \cdot 4} = \frac{52}{100} \Rightarrow 52\%$$

B. Since 8 is not a factor of 100, first use division to change the fraction to a decimal and then change the decimal to a percent by moving the decimal point two places to the right.

```
    0.625                 0.6 2 5  ⇒  62.5%
8 ) 5.000
    48
    ──
    20
    16
    ──
    40
    40
    ──
     0
```

Write each fraction or mixed number as a percent.

7. $\dfrac{17}{20}$ 　　　　**8.** $\dfrac{5}{12}$ 　　　　**9.** $1\dfrac{4}{5}$

To write a percent as a decimal, move the decimal point two places to the left and remove the percent symbol. To write a percent as a fraction, write the percent (without the percent sign) over 100 and reduce the fraction to lowest terms, if possible.

(continued)

Topic 3

Fractions, Decimals, and Percents

EXAMPLE 4

Write each percent as a decimal and as a fraction or mixed number.

    A. 46%   **B.** 175%

SOLUTION

    A. $46\% \Rightarrow 4\,6.\% \Rightarrow 0.46$ **B.** $175\% \Rightarrow 1\,7\,5.\% \Rightarrow 1.75$

        $46\% = \dfrac{46}{100} = \dfrac{23}{50}$ $175\% = \dfrac{175}{100} = \dfrac{7}{4} = 1\dfrac{3}{4}$

Write each percent as a decimal and as a fraction or mixed number.

10. 65%   **11.** 220%   **12.** 0.5%

To write a decimal as a fraction, write the number without the decimal point as the numerator of a fraction. The denominator of the fraction is the place value of the decimal digit that is farthest to the right. Then, write the fraction in lowest terms. For example, $0.5 = \frac{5}{10} = \frac{1}{2}$. To write a decimal as a percent, move the decimal point two places to the right and add a percent symbol.

EXAMPLE 5

Write 0.25 as (A) a percent and (B) a fraction in lowest terms.

SOLUTION

    A. $0.25 = 25\%$ Move decimal point two places to the right.

    B. $0.25 = \dfrac{25}{100}$ 25 hundredths can be written as $\frac{25}{100}$.

       $= \dfrac{1}{4}$ Write the fraction in lowest terms.

Write each decimal as a percent and as a fraction or mixed number in lowest terms.

13. 0.5   **14.** 0.185   **15.** 1.32

Mixed Review

16. Find the mean of the data 12, 8, 7, 16, 21, and 8.

17. Complete: A _____ graph is used to represent data as parts of a whole.

NAME _____ DATE _____

Quick Check

Review of Topic 3, Lesson 2

Standardized Testing Quick Check

1. There are 176 grams of fat in 16 tablespoons of mayonnaise. A recipe calls for 4 tablespoons of mayonnaise. How many grams of fat are in this much mayonnaise?

   A. 11 grams

   B. 88 grams

   C. 44 grams

   D. 704 grams

2. Tom drove 170 miles in $2\frac{5}{6}$ hours. If he keeps the same rate, how far will he have gone in 4 hours?

   A. 240 miles

   B. $\frac{1}{240}$ miles

   C. $1926\frac{2}{3}$ miles

   D. $120\frac{5}{12}$ miles

Homework Review Quick Check

3. Are the fractions $\frac{5}{9}$ and $\frac{2}{3}$ equivalent? Why or why not?

4. Find the missing number in the ratio: $\dfrac{14}{24} = \dfrac{?}{12}$.

Topic 3

Algebra 1

Basic Skills Workbook: Diagnosis and Remediation

NAME _____ DATE _____

Practice

For use with Lesson 3.3: Fractions, Decimals, and Percents

Write each fraction or mixed number as a decimal.

1. $\frac{6}{8}$
2. $\frac{7}{10}$
3. $\frac{1}{3}$
4. $\frac{13}{30}$

5. $8\frac{1}{4}$
6. $\frac{9}{16}$
7. $5\frac{37}{50}$
8. $\frac{3}{11}$

Write each fraction or mixed number in lowest terms.

9. $6\frac{7}{14}$
10. $\frac{3}{18}$
11. $\frac{15}{27}$
12. $\frac{72}{100}$

13. $\frac{13}{26}$
14. $5\frac{12}{15}$
15. $7\frac{8}{52}$
16. $2\frac{24}{36}$

Write each fraction or mixed number as a percent.

17. $\frac{4}{5}$
18. $\frac{1}{8}$
19. $\frac{17}{20}$
20. $\frac{5}{2}$

21. $\frac{9}{10}$
22. $5\frac{3}{4}$
23. $\frac{45}{50}$
24. $\frac{1}{3}$

25. $\frac{15}{40}$
26. $1\frac{2}{5}$
27. $2\frac{1}{25}$
28. $7\frac{49}{50}$

Write each percent as a decimal.

29. 51%
30. $52.\overline{3}\%$
31. 102%
32. $2\frac{1}{2}\%$

33. $\frac{3}{4}\%$
34. 0.1%
35. 9%
36. 234%

Write each percent as a fraction or mixed number in lowest terms.

37. 99%
38. 125%
39. 50%
40. 3%

41. $\frac{3}{4}\%$
42. $12\frac{1}{2}\%$
43. 150%
44. 8%

45. 2%
46. 4%
47. 225%
48. 120%

Write each decimal as a percent and as a fraction or mixed number in lowest terms.

49. 0.25
50. 0.675
51. 1.2
52. 0.128

53. 0.06
54. 0.375
55. 0.76
56. 5.6

57. 0.44
58. 4.75
59. 0.001
60. 1.01

Algebra 1
Basic Skills Workbook: Diagnosis and Remediation

Topic 3

NAME _____ DATE _____

Finding a Percent of a Number

GOAL Use decimal multiplication or fractions to find a percent of a number.

> Most states have a sales tax, expressed as a percent. For example, if the sales tax rate is 6%, this means that for each dollar you pay, you must pay an extra six cents in tax. To find how much tax an item has, you need to find 6% of the amount of the item.

To find a percent of a given number, first write the percent as a decimal or as a fraction. Then multiply.

EXAMPLE 1 ───

    A. What is 15% of 60?   **B.** What is 25% of 48?

SOLUTION

    A. $15\% = 0.15$ Write 15% as a decimal.

         $0.15 \times 60 = 9$ Multiply.

         15% of 60 is 9.

    B. $25\% = \dfrac{1}{4}$ Write 25% as a fraction.

         $\dfrac{1}{4} \times 48 = 12$ Multiply.

         25% of 48 is 12.

Find the percent of the number.

1. 50% of 274   **2.** 75% of 836   **3.** 14% of 1203

4. 39% of 523.6   **5.** 0.5% of 123   **6.** 95.6% of 95.6

You can solve many real-life problems by finding a percent of a number.

EXAMPLE 2 ───

The regular price of a video game is $54.90. It is on sale for 30% off the regular price. What is the discount? What is the sale price?

SOLUTION

     Discount = Percent off $\times$ Regular Price

          $= 30\% \times 54.90$ Substitute.

          $= 0.3 \times 54.90$ Rewrite 30% as 0.3.

          $= \$16.47$ Multiply.

The discount is $16.47. To find the sale price, subtract the discount from the regular price. So $54.90 − $16.47 = $38.43. The sale price is $38.43.

(continued)

NAME _____ DATE _____

Finding a Percent of a Number

7. The regular price of a VCR is $179.90. It is on sale for 15% off the regular price. How much is the discount? What is the sale price of the VCR?

8. The sales tax on the VCR in Exercise 7 is 5.25%. What is the amount of sales tax on the discounted VCR? What is the total sale price of the VCR?

Percents are also used to help you analyze data. For example, you can read a bar graph that shows people's favorite types of food.

EXAMPLE 3

In a survey, 500 people were asked to name their favorite types of food. The five most popular are shown in the bar graph. Find the number of people who named each type of food.

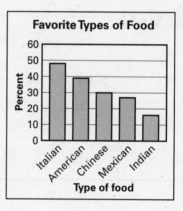

SOLUTION

There are 500 people in the survey. Multiply the percent by 500 to find the number of people who named each type of food.

Type of Food	Percent	Number who named food
Italian	48%	$0.48 \times 500 = 240$
American	39%	$0.39 \times 500 = 195$
Chinese	30%	$0.30 \times 500 = 150$
Mexican	27%	$0.27 \times 500 = 135$
Indian	16%	$0.16 \times 500 = 80$

Given the percent of households, find the number of surveyed households with each type of pet. About 80,000 households in the United States were surveyed.

9. Dog: 32%   **10.** Cat: 27%   **11.** Bird: 4.6%

Mixed Review

12. Write the ratio of 1 in. to 20 feet in lowest terms.

13. Write the fraction $\frac{24}{25}$ as a decimal and as a percent.

Algebra 1
Basic Skills Workbook: Diagnosis and Remediation

NAME _____ DATE _____

Quick Check

Review of Topic 3, Lesson 3

Standardized Testing Quick Check

1. Which decimal is equal to the fraction $\frac{3}{5}$?

    A. 0.15

    B. 0.6

    C. 0.35

    D. 1.67

2. Which fraction is equal to the decimal 0.375?

    A. $\frac{2}{3}$

    B. $\frac{3}{8}$

    C. $\frac{75}{2}$

    D. $\frac{3}{75}$

Homework Review Quick Check

Write each percent as a fraction or mixed number in lowest terms.

3. 175%   4. 46%   5. 16%   6. 28%

Write each fraction or mixed number as a percent.

7. $\frac{4}{11}$ 8. $\frac{13}{12}$ 9. $6\frac{4}{5}$ 10. $\frac{1}{9}$

Topic 3

Algebra 1
Basic Skills Workbook: Diagnosis and Remediation

Practice

For use with Lesson 3.4: Find a Percent of a Number

Find each number.

1. 30% of 80

2. 20% of 30

3. 15% of 240

4. 85% of 300

5. 55% of 125

6. 120% of 42

7. 70% of 28

8. 62% of 108

9. 83% of 150

10. 47% of 356

11. 165% of 684

12. 0.5% of 19

13. 25% of 25.99

14. 33% of 6574

15. 85% of 186.3

16. 112% of 487.6

17. If the sales tax rate is 6.5%, how much tax will you pay for an item whose price is $20?

18. The regular price of a book is $16.40. A store is having a 20% off sale. How much is the discount? What is the sale price?

19. You have a collection of 123 books. If about 40% of these books are mysteries, how many mysteries do you have in your collection?

20. You work in a restaurant for $5.75 per hour. You are given an 8% raise. What is your new hourly wage?

21. During the summer you mow lawns and do yard work. Last summer you charged $6.75 per hour. This year you decide to raise your rate 16%. What is the new rate per hour?

In Exercises 22–26, use the table at the right. It shows the percent of urban land area and rural land area of the total land area for different states. Find the area of each state that is considered urban area and rural area.

State	Urban	Rural
Vermont	1.5%	98.5%
California	5.2%	94.8%
Delaware	10.7%	89.3%
New Jersey	32.7%	67.3%
Nevada	0.9%	99.1%

22. Vermont   9,249.3 square miles

23. California   155,973.2 square miles

24. Delaware   1,954.6 square miles

25. New Jersey   7,418.8 square miles

26. Nevada   109,805.5 square miles

Topic 3

Algebra 1
Basic Skills Workbook: Diagnosis and Remediation

NAME _____ DATE _____

Assessment

For use with Topic 3: Rates, Ratios, and Percents

Write each fraction or mixed number as a decimal and as a percent.

1. $\dfrac{43}{50}$ 　　　　**2.** $3\dfrac{3}{5}$ 　　　　**3.** $\dfrac{3}{8}$

Write each fraction or mixed number in lowest terms.

4. $\dfrac{27}{45}$ 　　　　**5.** $\dfrac{24}{56}$ 　　　　**6.** $7\dfrac{15}{75}$

Write each percent as a decimal and as a fraction or mixed number in lowest terms.

7. 96% 　　　　**8.** 110% 　　　　**9.** 6%

Students were asked to choose which they prefer, orange juice or grapefruit juice. The results are shown in the table below.

Group	Prefer orange juice	Prefer grapefruit juice
Girls	50	18
Boys	48	10

Write each ratio in lowest terms.

10. girls preferring orange juice to boys preferring orange juice

11. boys preferring orange juice to boys preferring grapefruit juice

Write each ratio as a fraction in lowest terms.

12. 3 minutes to 25 seconds 　　　　**13.** 2 weeks : 6 days

14. 5 hours to 20 minutes

Find the unit rate.

15. 192 points in 8 games 　　　　**16.** 330 miles on 12 gallons

17. $13 for 4 pounds 　　　　**18.** 141 pages in 3 chapters

Find the missing number.

19. $\dfrac{4}{7} = \dfrac{?}{28}$ 　　**20.** $\dfrac{18}{36} = \dfrac{?}{6}$ 　　**21.** $\dfrac{6}{8} = \dfrac{3}{?}$ 　　**22.** $\dfrac{16}{?} = \dfrac{32}{18}$

Find each answer.

23. What number is 12% of 120? 　　**24.** 2% of 20 is what number?

25. 35% of 70 is what number? 　　**26.** 300% of 64 is what number?

Basic Skills Workbook: Diagnosis and Remediation

Topic 3

NAME _____ DATE _____

Patterns in Geometry

GOAL **Make predictions based on patterns in geometric figures, perimeters, and so on.**

> You can sometimes solve problems by finding a pattern in the given information. This is often true with problems involving geometric figures.

Understanding the Main Ideas

Recognizing a pattern among geometric figures may help you to write a variable expression that describes the pattern. The expression can be used to make predictions and to solve problems.

EXAMPLE 1

Predict the number of small triangles in the tenth figure of the pattern at the right.

SOLUTION

First, count the number of small triangles in each figure shown and record them in a table.

Figure number	1	2	3	4
Number of small triangles	1	4	9	16

Now look for a pattern as you compare each figure number with its number of small triangles. Notice that in each case, the number of small triangles is the square of the figure number. You can describe this pattern using a variable expression.

If the variable n represents the figure number, then

number of small triangles = square of figure number = n^2.

To find the number of small triangles in the tenth figure, use $n = 10$.

n^2 Write variable expression.

$10^2 = 100$ Substitute 10 for n and simplify.

Therefore, the number of small triangles in the tenth figure should be 100.

(continued)

Patterns in Geometry

1. The figures below model the first four *triangular numbers*.

Which expression represents the *n*th triangular number?

A. $2n - 1$ **B.** $3n - 2$ **C.** $\dfrac{n(n + 1)}{2}$ **D.** $n^2 - n$

2. Sketch the dot model for the eighth triangular number. Verify your model using your answer to Exercise 1.

3. How many dots are needed in the dot model representing the 15th triangular number? the 20th? the 30th?

Mixed Review

4. About 71% of the surface of Earth is covered by water. If the total surface area of the planet is about 197 million square miles, find the surface area covered by water.

5. Write the unit rate: It costs $18 for 4 tickets.

Topic 4

NAME _____ DATE _____

Quick Check

Review of Topic 3, Lesson 4

Standardized Testing Quick Check

1. The cost of Minh's restaurant meal is $9.75. If Minh plans to tip the server 20% of that amount, which is the best estimate of the tip?

   A. $1.00

   B. $1.25

   C. $1.50

   D. $2.00

2. From Exercise 1, how much did Minh spend at the restaurant, including the tip?

   A. $7.80

   B. $11.70

   C. $1.95

   D. $10.75

Homework Review Quick Check

3. 20% of 85 is what number?

4. 0.6% of 225 is what number?

5. An advertisement claimed that 80% of the dentists surveyed preferred Stainex Toothpaste over every other leading brand. If only 15 dentists took part in the survey, how many preferred Stainex?

Algebra 1
Basic Skills Workbook: Diagnosis and Remediation

NAME _____ DATE _____

Practice

For use with Lesson 4.1: Patterns in Geometry

The figures at the right are models for the first four *oblong numbers*.

1. Complete the table below.

Figure number	1	2	3	4
Oblong number	2			

2. Write the next three oblong numbers.

3. Write a variable expression that represents the *n*th oblong number.

4. Use your answer for Exercise 3 to find the 15th oblong number.

Use the figures at the right for Exercises 5–9.

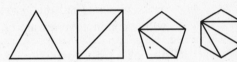

5. Complete the table below.

Number of sides	3	4	5	6
Number of triangles	1			

6. How is the number of triangles related to the number of sides?

7. Sketch the next two figures. What is the number of triangles for these two figures?

8. Write a variable expression for the number of triangles when the number of sides is *n*.

9. How many triangles are there when the figure has 20 sides?

The figures at the right show the number of line segments that can be drawn connecting a number of points when no three of the points lie on the same line.

10. Draw the next two figures in the pattern. Be sure that no three of the points you draw lie on the same line.

11. Complete the table below.

Number of points	2	3	4	5	6	7
Number of line segments	1					

12. Write a variable expression that represents the number of line segments that can be drawn connecting *n* points, no three of which lie on the same line.

Algebra 1
Basic Skills Workbook: Diagnosis and Remediation

Topic 4

NAME _____ DATE _____

Polygons

GOAL Identify polygons and their parts. Sketch polygons.

> In this lesson, you will study geometric figures called *polygons*. Special attention will be given to three types of polygons: rectangles, squares, and parallelograms.

Terms to Know

Example/Illustration

Terms to Know	Example/Illustration
Polygon a closed plane figure formed by line segments (called *sides*), each intersecting exactly two other sides, one at each endpoint (called a *vertex*) (*Note:* No two sides with a common vertex are on the same line.)	
Sides of a polygon the line segments that form a polygon	side
Vertex of a polygon a point where two sides of the polygon meet (The plural of vertex is *vertices*.)	vertex
Diagonal of a polygon a line segment that joins two nonconsecutive vertices of the polygon	diagonal
Congruent having the same size and shape (Congruent segments have the same length; congruent angles have equal measures.)	A B C In the figure, the tick marks indicate that $\overline{AB}$ and $\overline{AC}$ are congruent. The arcs indicate that $\angle B$ and $\angle C$ are congruent.
Regular polygon a polygon in which all of the sides are congruent and all of the angles are congruent	

(continued)

NAME _____ DATE _____

Polygons

Understanding the Main Ideas

A polygon is identified by the number of sides it has. The table below shows the names of several polygons.

Name	Number of sides	Name	Number of sides
Triangle	3	Hexagon	6
Quadrilateral	4	Heptagon	7
Pentagon	5	Octagon	8

Some polygons have special names. For example, three of the special quadrilaterals and their special properties are given below. (*Note:* The arrowheads in the figures indicate pairs of parallel lines.)

Parallelogram
two pairs of parallel sides; two pairs of congruent sides

Rectangle
four right angles; two pairs of congruent parallel sides

Square
four right angles; four congruent sides; two pairs of parallel sides

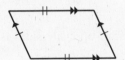   

Notice that every square is also a rectangle and that every rectangle is a parallelogram.

EXAMPLE 1

Identify the polygon shown at the right. Then name its sides, its vertices, and its diagonals.

SOLUTION

The polygon has four sides, so it is a quadrilateral. A polygon is named by listing its vertices in order as you go around the figure. So one possible name for a polygon is quadrilateral *ABCD*.

The sides of the quadrilateral are $\overline{AB}$, $\overline{BC}$, $\overline{CD}$, and $\overline{AD}$. Its vertices are *A*, *B*, *C*, and *D*. Its two diagonals (shown in the figure at the right) are $\overline{AC}$ and $\overline{BD}$.

(continued)

NAME _____ DATE _____

Polygons

1. Explain why each figure is *not* a polygon.

   a.

   b.

2. Identify the polygon at the right. Then name its sides, its vertices, and its diagonals.

EXAMPLE 2

Sketch a quadrilateral that is not a parallelogram.

SOLUTION

The figure must have four sides, and at least one pair of opposite sides is not parallel. Three possible figures are shown below.

 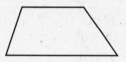

3. Sketch a hexagon that is not regular.

4. Sketch an octagon that is regular.

Mixed Review

5. Solve the proportion $\frac{3}{8} = \frac{x}{20}$ for x.

6. Write a variable expression that represents the nth number in the pattern 3, 6, 9, 12, 15, ...

Algebra 1
Basic Skills Workbook: Diagnosis and Remediation

NAME _____ DATE _____

Quick Check

Review of Topic 4, Lesson 1

Standardized Testing Quick Check

1. A pie is to be sliced into equal servings by cutting along lines through the center of the pie. Predict how many cuts must be made to result in 12 equal servings.

1 cut   2 cuts   3 cuts

   A. 4 cuts   **B.** 6 cuts   **C.** 8 cuts   **D.** 16 cuts   **E.** None of these

2. Choose the variable expression that represents the value of the *n*th term of the pattern.

Term number	1	2	3	4	5
Value of term	$\frac{1}{2}$	2	$\frac{9}{2}$	8	$\frac{25}{2}$

   A. $\dfrac{n}{2}$

   B. n

   C. n^2

   D. $\dfrac{n^2}{2}$

Homework Review Quick Check

3. Write an expression for the value of the *n*th term of the pattern.

Term number	1	2	3	4	5
Value	3	5	7	9	11

Tell whether each figure is a polygon. If it is not, explain why it is not.

1.

2.

3.

Complete each statement.

4. A pentagon is a polygon with _____ sides.

5. A regular hexagon has six sides that are _____ .

6. A polygon has eight congruent sides and eight congruent angles. The polygon is a _____ octagon.

Identify each polygon. Be as specific as possible.

7.

8.

9.

Name each polygon in at least three ways. Then name its sides, its vertices, and its diagonals.

10.

11.

12.

13.

14.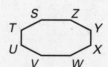

15.

Sketch each polygon, if possible. If it is not possible, explain why it is not.

16. a rectangle that is not a square

17. a pentagon

18. a square that is not a rectangle

19. a quadrilateral that includes exactly one right angle

20. a parallelogram that is not a rectangle

21. a triangle that is not regular

Topic 4

NAME _____ DATE _____

Perimeters and Areas of Polygons

GOAL **Find perimeters and areas of triangles, rectangles, squares, and parallelograms.**

> The *perimeter* of a polygon is the sum of the lengths of all its sides. The *area* of a polygon is the number of square units enclosed by the polygon. It is helpful to use formulas for finding the area.

Terms to Know	**Example/Illustration**
Bases of a parallelogram either pair of parallel sides of a parallelogram	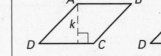 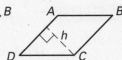
Height of a parallelogram the perpendicular distance between the bases of a parallelogram	If $\overline{AB}$ and $\overline{DC}$ are considered the bases of the parallelogram, then k is the height. If $\overline{AD}$ and $\overline{BC}$ are considered the bases of the parallelogram, then h is the height.
Height of a triangle the perpendicular distance between the base of the triangle and the vertex opposite that base (*Note:* Any side of a triangle may be considered its base.)	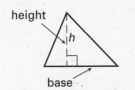

Understanding the Main Ideas

You are probably familiar with the formula for the area of a rectangle, $A = \ell w$. The formulas for the area of a square, a triangle, and a parallelogram can all be derived from that formula. The perimeter of each figure can be found by adding the lengths of all the sides.

Rectangle
The area of a rectangle is the length times the width.

$A = \ell w$

Square
The area of a square is the square of the length of a side.

$A = s^2$

(continued)

Algebra 1 **75**
Basic Skills Workbook: Diagnosis and Remediation

Topic 4

NAME _____ DATE _____

Perimeters and Areas of Polygons

Parallelogram

The area of a parallelogram is the base times the height.

$A = bh$

Triangle

The area of a triangle is one-half the base times the height.

$A = \frac{1}{2}bh$

EXAMPLE 1 ───────────────────────────────────

Find the area and perimeter of a rectangle with length 3 cm and height 2 cm.

SOLUTION

$A = \ell w = 3(2)$, or 6

The opposite sides of a rectangle are congruent. Therefore,

$$P = 2\ell + 2w$$
$$= 2(3) + 2(2), \text{ or } 10.$$

The area of the rectangle is 6 cm² and the perimeter is 10 cm.

EXAMPLE 2 ───────────────────────────────────

Find the area and perimeter of a square with 5 in. sides.

SOLUTION

$A = s^2 = 5^2$, or 25

All four sides of a square are congruent. Therefore,

$$P = 4s$$
$$= 4(5), \text{ or } 20.$$

The area of the square is 25 in.² and the perimeter is 20 in.

EXAMPLE 3 ───────────────────────────────────

Find the area and perimeter of the parallelogram.

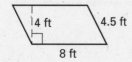

(continued)

Topic 4

NAME _____ DATE _____

Perimeters and Areas of Polygons

SOLUTION

$A = bh = 8(4)$, or 32   The height is 4, not 4.5!

The opposite sides of a parallelogram are congruent. Therefore,

$$P = 2(4.5) + 2(8)$$
$$= 9 + 16, \text{ or } 25.$$

The area of the parallelogram is 32 ft² and the perimeter is 25 ft.

EXAMPLE 4 _____

Find the area and perimeter of the triangle.

SOLUTION

$A = \frac{1}{2}bh = \frac{1}{2}(4)(3)$, or 6 $P = 3 + 4 + 5$, or 12

The area of the triangle is 6 m² and the perimeter is 12 m.

Find the area and perimeter of each figure.

1.

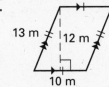

2.

3.

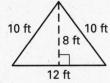

4.

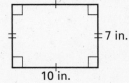

Mixed Review

5. One side of a parallelogram is 3 in. long. Another side is 4 in. long. How long are the other two sides?

6. The scale of a drawing of a house is 1 in. : $7\frac{1}{2}$ ft. Write the scale as a ratio in lowest terms.

Topic 4

Algebra 1
Basic Skills Workbook: Diagnosis and Remediation

Quick Check

Review of Topic 4, Lesson 2

Standardized Testing Quick Check

1. Polygon *ABCDEF* is a regular hexagon. Which statement is false?

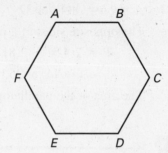

   A. $\angle B$ and $\angle F$ are congruent.

   B. $\overline{AB}$ and $\overline{EF}$ are congruent.

   C. If $AB = x$, then the area of hexagon *ABCDEF* is $6x^2$.

   D. If $CD = y$, then the perimeter of hexagon *ABCDEF* is $6y$.

Homework Review Quick Check

2. Use the figure at the right.

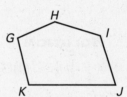

   A. Identify the polygon.

   B. Is the polygon regular?

   C. Give two names for the polygon.

   D. Name the sides of the polygon.

   E. Name the vertices of the polygon.

   F. Name the diagonals of the polygon.

3. Sketch a regular quadrilateral.

Topic 4

LESSON 4.3

NAME _____ DATE _____

Practice

For use with Lesson 4.3: Perimeters and Areas of Polygons

Find the perimeter of each figure.

1.

15 ft

24 ft

2.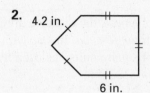

4.2 in.

6 in.

3.

19 cm

4.

2 in.

4 in.

5 in.

8 in.

5.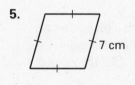

7 cm

6.

13 m

5 m

For Exercises 7–12, find the area of each polygon.

7.

12.5 in.

8.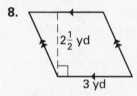

$2\frac{1}{2}$ yd

3 yd

9.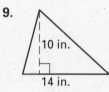

10 in.

14 in.

10.

21 km

15 km

11.

15 mm

10 mm

12.

10 m   8 m

13. One side of a regular hexagon is 13 cm long. Find the perimeter of the hexagon.

14. A parallelogram has base 18 m long and height of 9 m. Find the area of the parallelogram.

15. A triangle has base 7 yd long and height 4 yd. Find the area of the triangle.

16. A square garden is 30 ft long on a side. Find the perimeter and area of the garden.

NAME _____ DATE _____

Circles and Areas

GOAL Find the area and circumference of a circle.

> The perimeter of a polygon is the distance around it. The distance around a circle is called its *circumference*. Just as with a polygon, the area of a circle is the number of square units enclosed by it.

Terms to Know

Example/Illustration

Circle
the set of all points in a plane that are a given distance (called the *radius*) from a given point (called the *center*) in the plane
Radius of a circle
any line segment with one endpoint at the center of the circle and the other endpoint on the circle; also, the length of such a segment
Diameter of a circle
a line segment that passes through the center of the circle and has both endpoints on the circle; also, the length of such a segment

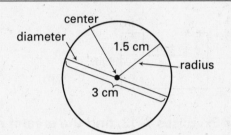

The radius *r* of the circle is 1.5 cm.
The diameter *d* of the circle is 3 cm.

Understanding the Main Ideas

The ratio of the circumference C of a circle to its diameter d is the same regardless of the size of the circle. The ratio $\dfrac{C}{d}$ is equal to the number π (the Greek letter "pi"). The exact value of π cannot be expressed as a fraction or as a finite decimal. However, it can be *approximated* by the fraction $\dfrac{22}{7}$ or the decimal 3.14.

Whichever approximation is more convenient may be used. Since $\dfrac{C}{d} = \pi$ for all circles, the circumference of a circle can be found using the formula $C = \pi d$. Since the diameter of a circle is twice the radius of the circle, the formula can be rewritten as $C = 2\pi r$.

(continued)

Topic 4

NAME _____ DATE _____

Circles and Areas

EXAMPLE 1 ────────────────────────────────────

The diameter of a circle is 14 cm. Find the circumference of the circle. Use $\frac{22}{7}$ for π.

SOLUTION

$C = \pi d$

$C \approx \frac{22}{7}(14)$

$\quad = 44$

The circumference of the circle is about 44 cm.

Find the circumference of each circle.

1. Use 3.14 for π.

8 cm

2. Use $\frac{22}{7}$ for π.

28 in.

The formula for the area A of a circle with radius r is $A = \pi r^2$.

EXAMPLE 2 ────────────────────────────────────

The diameter of a circle is 12 in. Find the area of the circle.

SOLUTION

Since the diameter of the circle is 12 in. and the diameter of a circle is twice its radius, the radius of the circle is 6 in. (*Note:* Since 6^2 is not a multiple of 7, we will use the approximation 3.14 for π.)

$A = \pi r^2$

$\quad \approx 3.14(6^2)$

$\quad = 3.14(36)$

$\quad = 113.04$

The area of the circle is about 113 in.^{2}.

(continued)

Algebra 1
Basic Skills Workbook: Diagnosis and Remediation

81

Topic 4

NAME _____ DATE _____

Circles and Areas

Find the area of each circle.

3. Use 3.14 for π.

10 yd

4. Use $\frac{22}{7}$ for π.

7 m

Note: If you use an approximate value of π when calculating the circumference or area of a circle, your answers are approximations. If you want to give an exact value for the circumference or area of a circle, simply leave your answer in terms of π. For example, the exact circumference of the circle in Example 1 is 14π cm and the exact area of the circle in Example 2 is 36π in.^{2}.

Mixed Review

5. A rectangular house lot is 150 ft long and 100 ft wide. Find the area of the lot.

6. Which is greater, the area of a triangle with base 16 cm long and height 8 cm, or the area of a square with sides 8 cm long?

7. The circle graph at the right shows the breakdown of income for Music Mania last year.

    a. What type of sales accounted for more than $\frac{3}{4}$ of the income?

    b. The total income for the year was $100,000. About how many dollars of income were from the sale of CDs?

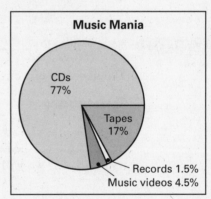

Music Mania

CDs 77%

Tapes 17%

Records 1.5%
Music videos 4.5%

Algebra 1
Basic Skills Workbook: Diagnosis and Remediation

Topic 4

NAME _____ DATE _____

Quick Check

Review of Topic 4, Lesson 3

Standardized Testing Quick Check

1. A triangle is 16 cm high and its base is 12 cm long.
   Find the area of the triangle.

   A. 28 cm²

   B. 64 cm²

   C. 96 cm²

   D. 192 cm²

Homework Review Quick Check

Find the perimeter of each polygon.

2.
7 ft 5 ft 6 ft

3.
6.5 cm

Find the area of each polygon.

4.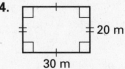
20 m
30 m

5.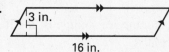
3 in.
16 in.

NAME _____ DATE _____

Practice

For use with Lesson 4.4: Circles and Area

Throughout these exercises, use either $\frac{22}{7}$ or 3.14 for π. You may also use a calculator. Round answers to the nearest hundredth when necessary. Find the circumference of the circle with the given diameter or radius.

1. $r = 20$ m

2. $d = 82$ ft

3. $d = 56$ cm

4. $r = 12$ in.

5. $d = 6$ yd

6. $r = 60$ m

7. $r = 140$ mm

8. $d = 100$ m

9. $d = 52$ ft

10. $r = 35$ ft

For Exercises 11–20, find the area of the circle with the given diameter or radius.

11. $r = 7$ cm

12. $r = 1\frac{1}{2}$ ft

13. $d = 20$ in.

14. $d = 21$ yd

15. $r = 3.5$ m

16. $r = 3$ cm

17. $d = 9$ ft

18. $d = 16$ m

19. $d = 18$ yd

20. $r = 70$ in.

21. The circumference of a circle is 264 ft. Find the radius and the diameter of the circle.

22. The area of a circle is 314 m². Find the radius and the diameter of the circle.

23. A circular rug is 10 ft in diameter. A strip of fringe is to be added around the edge of the rug. How long must the strip be?

24. Find the area of the shaded region. Use 3.14 for π.

8 cm

NAME _____ DATE _____

Assessment

For use with Topic 4: Geometry

The first four figures in a pattern of triangles are shown below. The length of each side of each small triangle is 1 unit.

1. Find the perimeter of each figure in the pattern.

2. Write a variable expression for the perimeter of the *n*th figure.

3. Use your answer for Exercise 2 to find the perimeter of the 20th figure.

Sketch each polygon.

4. a rectangle whose length is twice its width

5. a parallelogram with four congruent sides

6. a pentagon that is not regular

Tell whether each statement about the polygon shown at the right is *True* or *False*.

7. One correct name for the polygon is *STUVRW*.

8. The polygon has six vertices.

9. $\overline{RW}$ is a diagonal of the polygon.

10. The polygon is not regular.

11. The polygon has exactly three diagonals.

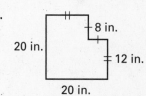

Find the perimeter of each polygon.

12.

15 cm
8 cm

13.
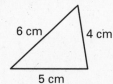
6 cm 4 cm
5 cm

14.
8 in.
20 in.
12 in.
20 in.

Find the area of each polygon.

15.

15 ft
21 ft

16.

30 mm
24 mm

17.
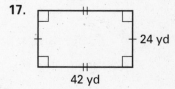
24 yd
42 yd

(continued)

NAME _____ DATE _____

Assessment

For use with Topic 4: Geometry

Find the circumference of the circle with the given radius or diameter. Use 3.14 or $\frac{22}{7}$ for π.

18. $r = 50$ cm

19. $d = 22$ in.

20. $r = 3\frac{1}{2}$ yd

Find the area of the circle with the given radius or diameter.

21. $r = 17$ ft

22. $d = 40$ m

23. $d = 140$ mm

Topic 4

Integer Concepts

GOAL **Graph integers, find opposites, and find absolute values of integers using a number line.**

> Many real-life situations can be modeled with whole numbers. Some measures, such as temperatures, are more easily modeled with an expanded set of numbers called *integers*. For example, a thermometer can be thought of as a vertical number line where, on the Celsius scale, negative temperatures are below freezing.

Terms to Know	*Example/Illustration*
Integers any number in the list $\ldots, -3, -2, -1, 0, 1, 2, 3, \ldots$ (On a horizontal number line, the negative integers are to the left of 0 and the positive integers are to the right of 0.)	$\ldots -4, -3, -2, -1, 0, 1, 2, 3, 4, \ldots$ Negative integers, Zero, Positive integers −3 −2 −1 0 1 2 3
Opposites two numbers that are the same distance from 0 on a number line, but are on opposite sides of 0. (The opposite of 0 is 0.)	opposites −3 −2 −1 0 1 2 3 −2 and 2 are each 2 units from 0, so −2 and 2 are opposites.
Absolute value on a number line, the distance from the number to 0 (The symbol $|x|$ is read "the absolute value of x.")	⊢2 units⊢2 units⊣ −3 −2 −1 0 1 2 3 $|-2| = 2$ and $|2| = 2$

Understanding the Main Ideas

To understand integers, you must first be able to graph them on a number line, as shown in the following example.

EXAMPLE 1 _____

Graph $-3, -1, 0, 2$ and 3 on a number line. Name the two integers that are opposites.

SOLUTION

Step 1: Use a ruler to draw a straight line.

Step 2: Mark several evenly spaced tick marks on the line and label them. Draw arrowheads on the ends of the line.

(continued)

NAME _____ DATE _____

Integer Concepts

Step 3: Graph each number on the number line.

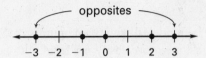

Because -3 and 3 are the same distance from 0, they are opposites.

It is helpful to know how to put integers in order from least to greatest. For example, you may be asked to analyze scientific data that involves negative integers. Putting data in order from least to greatest may help you understand the data better.

EXAMPLE 2

A. Name the integers between -8 and -1.

B. Order $-5, 4, -6, 2, 0,$ and -3 from least to greatest.

SOLUTION

A. The integers between -8 and -1 are $-7, -6, -5, -4, -3,$ and -2.

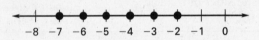

B. To order $-5, 4, -6, 2, 0,$ and -3 from least to greatest, begin by graphing the numbers on a number line.

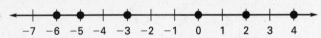

To order the integers from least to greatest, read from left to right. The order is $-6, -5, -3, 0, 2, 4$.

Order the integers from least to greatest.

1. $-6, 4, 3, -5, -1, 0$ **2.** $-8, -6, -3, -9, -4, -10$

3. $10, 16, -20, -18, -25, 30$

EXAMPLE 3

Find the absolute value of each number.

A. -2 **B.** 4

(continued)

Algebra 1
Basic Skills Workbook: Diagnosis and Remediation

NAME _____ DATE _____

Integer Concepts

SOLUTION

You can use a number line to find the absolute value.

A. $|-2| = 2$ Absolute value of -2 is 2.

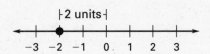

B. $|4| = 4$ Absolute value of 4 is 4.

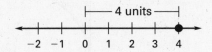

Graph the integer on a number line. Then find the absolute value of the integer.

4. 3   **5.** -1 **6.** 0   **7.** -15

Mixed Review

8. Identify the polygon at the right. Then name its sides, its vertices, and its diagonals.

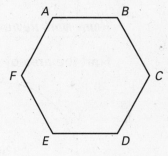

LESSON

5.1

Quick Check

Review of Topic 4, Lesson 4

Standardized Testing Quick Check

1. A circle has a circumference of 226.08 centimeters. Which value best approximates the circle's radius?

   A. 36 cm

   B. 226 cm

   C. 72 cm

   D. 1017 cm

2. Estimate the area of the shaded region.

   A. 17.15 cm²

   B. 23.43 cm²

   C. 7.74 cm²

   D. 6.75 cm²

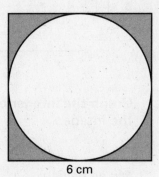

6 cm

Homework Review Quick Check

Find the area of each circle. Use 3.14 or $\frac{22}{7}$ for π.

3.

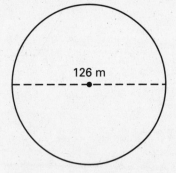

126 m

4.

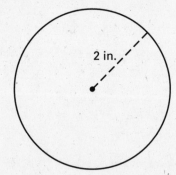

2 in.

5. *Flowers*   You want to plant flowers around a circular water fountain in your yard. The diameter of the fountain is 20 feet. You want to plant a flower no closer than every 2 feet. What is the maximum number of flowers you can plant around the fountain?

6. *Radio Waves*   A radio station broadcasts a signal that extends in a 75-mile radius from the station. Find the area of the region that receives the radio station's broadcasts.

Algebra 1
Basic Skills Workbook: Diagnosis and Remediation

LESSON 5.1

NAME _____ DATE _____

Practice

For use with Lesson 5.1: Integer Concepts

Graph the integers on a number line.

1. $-4, -1, 3, -7, 7$ **2.** $10, -2, -6, 1, 0$ **3.** $0, 5, -3, 9, -8$

4. $12, -12, -15, 20, 10$ **5.** $-6, -8, -4, 8, 4$ **6.** $-10, -5, -20, -18, -13$

Name the opposite of the number.

7. -9 **8.** 15 **9.** 28

10. -4 **11.** -17 **12.** 100

Order the integers from least to greatest.

13. $15, 25, -31, -22, 36, -46$ **14.** $-103, -105, -100, -110, -101$

15. $85, 60, 77, 100, 30, 55$ **16.** $66, -3, -10, 39, 51, -25$

17. $116, 106, -106, -110, 110$ **18.** $-79, -1110, -80, 85, 32$

Find the absolute value.

19. $|21|$ **20.** $|-11|$ **21.** $|34|$

22. $|-34|$ **23.** $|15|$ **24.** $|-4|$

25. $|-19|$ **26.** $|0|$ **27.** $|13|$

Complete the statement using $<$, $>$, or $=$.

28. 4 ▢ $|-4|$ **29.** -7 ▢ $|-7|$ **30.** $|-1|$ ▢ 1

31. $|-18|$ ▢ $|10|$ **32.** $|-16|$ ▢ 0 **33.** $|-22|$ ▢ $|-23|$

In Exercises 34–38, use the number line below.

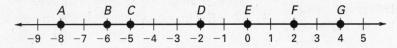

34. Which labeled point has an absolute value of 6?

35. Which labeled point is the same distance from 0 as -4?

36. Which labeled point is the opposite of 5?

37. Which labeled point is its own absolute value?

38. Name two labeled points that have the same absolute value.

NAME _____ DATE _____

Adding Integers

GOAL Add integers.

> Number lines not only show the relationships between positive and
> negative numbers, but they are very helpful in performing operations
> with positive and negative numbers.

Understanding the Main Ideas

You can draw arrows on a number line to model the addition of two integers. As
shown in the following example, an arrow pointing to the left on the number line
models a negative number, while an arrow pointing to the right models a positive
number.

EXAMPLE 1

Use a number line to find each sum.

A. $-1 + (-3)$ **B.** $6 + (-3)$

SOLUTION

A.

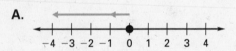

Start at 0 and draw an arrow 1 unit to
the left to model -1. From the tip of
this arrow, draw a second arrow 3
more units to the left to model adding
-3. The tip of this second arrow
models the sum: $-1 + (-3) = -4$.

B.

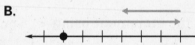

Start at 0 and draw an arrow 6 units to
the right to model 6. From the tip of this
arrow, draw a second arrow 3 units to
the left to model adding -3. The tip of
this second arrow models the sum:
$6 + (-3) = 3$.

Use a number line to find each sum.

1. $2 + 3$ **2.** $-3 + (-2)$ **3.** $-4 + (-2)$

When adding two integers with the same sign, follow these steps:

1. Find the absolute value of each number.

2. Add the two absolute values.

3. Give the result the same sign as the two numbers.

(continued)

NAME _____   DATE _____

Adding Integers

EXAMPLE 2

Find the sum $-7 + (-9)$.

SOLUTION

Step 1: Find the two absolute values: $|-7| = 7$; $|-9| = 9$.

Step 2: Add the two absolute values: $7 + 9 = 16$.

Step 3: Give the result the same sign as the numbers:
$-7 + (-9) = -16$.

Find each sum.

4. $11 + 19$ **5.** $-13 + (-20)$ **6.** $-9 + (-11)$

When adding two integers with different signs, follow these steps:

1. Find the absolute value of each number.

2. Subtract the lesser absolute value from the greater absolute value.

3. Give the result the same sign as the number with the greater absolute value.

EXAMPLE 3

Find the sum $-10 + 4$.

SOLUTION

Step 1: Find the two absolute values: $|-10| = 10$; $|4| = 4$.

Step 2: Subtract the lesser absolute value from the greater absolute value: $10 - 4 = 6$.

Step 3: Give the result the same sign as the number with the greater absolute value: $-10 + 4 = -6$.

Find each sum.

7. $-15 + 7$ **8.** $2 + (-11)$ **9.** $7 + (-13)$

Mixed Review

10. The diameter of a circular window is 18 in. Find the area of the window. Use 3.14 for π.

11. A rectangle is 18 km long and 9 km wide. Find the perimeter and area of the rectangle.

12. 85% of 200 is what number?

Algebra 1

Basic Skills Workbook: Diagnosis and Remediation

LESSON

5.2

NAME _____ DATE _____

Quick Check

Review of Topic 5, Lesson 1

Standardized Testing Quick Check

Temperatures   **In Exercises 1 and 2, use the following information.**

The table below shows the lowest temperatures on record for each month in Great Falls, Montana.

Month	J	F	M	A	M	J	J	A	S	O	N	D
Temperature (°F)	-37	-35	-29	-6	15	31	40	30	20	-11	-25	-43

1. Which month had the coldest temperature?

    A. November

    B. March

    C. July

    D. December

2. Which of the following temperatures are ordered from least to greatest?

    A. $15, 20, 30, 31, 40, -6, -11, -25, -29, -35, -37, -43$

    B. $-6, -11, -25, -29, -35, -37, -43, 15, 20, 30, 31, 40$

    C. $-43, -37, -35, -29, -25, -11, -6, 15, 20, 30, 31, 40$

    D. $40, 31, 30, 20, 15, -6, -11, -25, -29, -35, -37, -43$

Homework Review Quick Check

Find each absolute value.

3. $|6|$ **4.** $|0|$ **5.** $|-17|$ **6.** $|35|$

Algebra 1
Basic Skills Workbook: Diagnosis and Remediation

LESSON

5.2

NAME _____ DATE _____

Practice

For use with Lesson 5.2: Adding Integers

Topic 5

Complete the statement using the words *sometimes, always,* or *never*.

1. The sum of two negative numbers is _____ positive.

2. The sum of a positive number and a negative number is _____ negative.

3. The sum of a number and its opposite is _____ zero.

For Exercises 4–42, find each sum.

4. $-5 + (-5)$

5. $4 + 6$

6. $-5 + 3$

7. $7 + (-4)$

8. $-3 + 2$

9. $6 + (-7)$

10. $-43 + 51$

11. $17 + (-15)$

12. $-98 + 16$

13. $44 + (-61)$

14. $22 + 65$

15. $-19 + (-33)$

16. $-8 + (-6)$

17. $-26 + 13$

18. $111 + 33$

19. $88 + 12$

20. $50 + (-25)$

21. $-70 + (-8)$

22. $40 + (-29)$

23. $-12 + (-52)$

24. $-18 + 25$

25. $37 + (-19)$

26. $-29 + (-16)$

27. $71 + 54$

28. $-73 + 81$

29. $-10 + 42$

30. $28 + (-15)$

31. $-32 + 32$

32. $-40 + 88$

33. $95 + (-55)$

34. $18 + 87$

35. $-100 + 81$

36. $51 + 17$

37. $15 + (-9)$

38. $-21 + (-60)$

39. $76 + (-76)$

40. $-18 + 34$

41. $-31 + (-22)$

42. $75 + 29$

43. At 8 A.M., the temperature was $-5°F$. Over the next five hours, the temperature rose $16°F$. What was the temperature at 1 P.M.?

44. A football team lost 15 yards on one play. On the next play, they gained 8 yards. Find the total amount of yardage gained or lost on these two plays.

NAME _____ DATE _____

Subtracting Integers

GOAL Subtract integers.

> At 6:00 A.M. on the morning of January 12, 1911, the temperature in
> Rapid City, South Dakota, was 49°F. Over the next two hours, the
> temperature dropped 62°F! At 8:00 A.M., the temperature was −13°F.
> The temperature change can be represented by subtracting integers:
>
> $49 - 62 = -13.$

Understanding the Main Ideas

One way to subtract integers is to use integer chips. Of course, modeling the
subtraction $49 - 62$ would require a lot of integer chips, so we will start with a
simpler example.

EXAMPLE 1 _____

Find the difference $-5 - (-2)$.

SOLUTION

Start with 5 negative chips.   Remove 2 negative chips.   3 negative chips remain.

　　　　−5　　　　　　−　　　　　　−2　　　　=　　　　−3

Thus, $-5 - (-2) = -3.$

Find each difference.

1. $-8 - (-2)$ 　　　　　**2.** $-5 - (-1)$ 　　　　　**3.** $6 - 2$

When modeling the subtraction of integers with integer chips, many times the
chips that must be removed are not available. In order to have the required number of integer chips available, one or more *zero pairs* of chips must be added
to the model. A zero pair consists of one positive chip and one negative chip.
Since its value is $1 + (-1) = 0$, inserting such a pair does not change the value
of the model.

(continued)

NAME _____ DATE _____

Subtracting Integers

EXAMPLE 2

Find the difference $4 - 5$.

SOLUTION

Start with 4 positive chips.

In order to be able to remove 5 positive chips, one zero pair must first be added.

Now 5 positive chips can be removed.

The remaining chip models the difference.

Thus, $4 - 5 = -1$.

Find each difference.

4. $3 - 4$ **5.** $-2 - (-5)$ **6.** $2 - (-3)$

In the previous lesson, you saw that two numbers are *opposites* if their sum is 0. For example, since $3 + (-3) = 0$, 3 and -3 are opposites. Compare the following sum and difference involving opposites; they demonstrate a rule that can be used to subtract integers.

$$5 - 3 = 2 \qquad 5 + (-3) = 2$$

Subtracting Integers

Subtracting an integer is the same as adding its opposite. In general, for all integers a and b, $a - b = a + (-b)$.

EXAMPLE 3

Find the difference $14 - (-11)$.

SOLUTION

First, rewrite the subtraction as an addition. Use the fact that the opposite of -11 is 11. Then find the sum.

$$14 - (-11) = 14 + 11$$
$$= 25$$

(continued)

Subtracting Integers

Name the opposite of each number.

7. -9

8. 21

9. -1

Find each difference after rewriting it as a sum.

10. $8 - 10$

11. $12 - (-7)$

12. $-13 - 4$

Mixed Review

13. Identify the next two numbers in the pattern 11, 10, 8, 5, 1, . . .

14. A television weather reporter records the temperature every hour during a heat wave that lasts for 3 days. Which type of graph would be most appropriate for the reporter to display the data, a bar graph, a line graph, or a circle graph?

15. Find the area of a triangle with a base that is 15 cm long and a height of 12 cm.

16. You owe your sister $35. You earn $27 doing yard work. How much more money do you need in order to repay your sister the total amount you owe her?

NAME _____ DATE _____

Quick Check

Review of Topic 5, Lesson 2

Standardized Testing Quick Check

1. Tim is a marine biologist. He took a deep-sea dive in Hawaii and dove 1000 feet below sea level (-1000 feet). After he reached that depth, he rose back up 633 feet to investigate an interesting coral formation. At what depth was the coral formation?

   A. 1633 feet

   B. -367 feet

   C. -1633 feet

   D. 367 feet

Homework Review Quick Check

Find each sum.

   2. $-1.1 + (-39)$ **3.** $17 + 18$ **4.** $-21 + 14$

Algebra 1
Basic Skills Workbook: Diagnosis and Remediation

Practice

For use with Lesson 5.3: Subtracting Integers

Name the opposite of each number.

1. 35   **2.** -44 **3.** 0

4. -18 **5.** 71   **6.** -5

Write each subtraction as an addition.

7. $1 - (-15)$ **8.** $51 - 16$ **9.** $32 - (-19)$

10. $8 - 13$ **11.** $12 - (-17)$ **12.** $2 - 10$

For Exercises 13–48, find each difference.

13. $-11 - 8$ **14.** $-11 - (-8)$ **15.** $11 - 8$

16. $24 - 16$ **17.** $16 - 24$ **18.** $16 - (-24)$

19. $-8 - 22$ **20.** $31 - (-5)$ **21.** $-13 - 4$

22. $-7 - 21$ **23.** $44 - (-21)$ **24.** $-11 - (-23)$

25. $-8 - (-26)$ **26.** $15 - (-15)$ **27.** $-50 - 12$

28. $-15 - 9$ **29.** $4 - 17$ **30.** $-16 - (-5)$

31. $-30 - (-24)$ **32.** $7 - 17$ **33.** $-14 - 27$

34. $55 - (-21)$ **35.** $-45 - (-15)$ **36.** $19 - (-7)$

37. $-62 - 41$ **38.** $37 - 42$ **39.** $3 - 19$

40. $21 - 43$ **41.** $-1 - (-17)$ **42.** $-88 - 20$

43. $22 - (-17)$ **44.** $31 - 18$ **45.** $-18 - (-28)$

46. $37 - (-11)$ **47.** $25 - 34$ **48.** $-20 - (-4)$

49. In describing elevations, you can use negative numbers to represent elevations below sea level and positive numbers to represent elevations above sea level. The highest elevation in Louisiana is Driskill Mountain at 535 ft. This is 543 ft higher than the elevation of New Orleans.

    a. Find the elevation of New Orleans.

    b. How much lower than New Orleans is Death Valley, California, the lowest elevation in the United States, with an elevation of -282 ft?

NAME _____ DATE _____

Multiplying and Dividing Integers

GOAL **Multiply and divide integers.**

> When you add or subtract integers, the sign of the result depends on the
> absolute values of the integers. When you multiply or divide integers, the
> sign of the result depends only on the signs of the two integers.

Understanding the Main Ideas

In the following example, integer chips are used to show that the product of a
negative integer and a positive integer is negative.

EXAMPLE 1

Find the product $4(-3)$.

SOLUTION

Think of the product as 4 times a group of 3 negative chips, that is, 4 groups of
3 negative chips. When these groups are combined, there are 12 negative chips
representing the product -12.

4 groups of 3   1 group of 12
negative chips   negative chips

 →

$4(-3)$ = -12

Find each product.

1. $5(-2)$ **2.** $3(-3)$ **3.** $2(-4)$

Since the order in which two numbers are multiplied does not affect the product,
$4(-3) = (-3)(4)$. Therefore, the product of two integers with different signs is
always negative.

When you multiply two positive integers, you know that the product is positive.
But what happens when you multiply two negative integers?

(continued)

Algebra 1
Basic Skills Workbook: Diagnosis and Remediation

Multiplying and Dividing Integers

EXAMPLE 2

A. Find the products $3(-3)$, $2(-3)$, $1(-3)$, and $0(-3)$.

B. Identify the pattern in the products found in part (A) and use it to find the products $(-1)(-3)$, $(-2)(-3)$, and $(-3)(-3)$.

SOLUTION

A. Use the method of Example 1. Think of the products as 3 groups of 3 negative chips, 2 groups of 3 negative chips, 1 group of 3 negative chips, and 0 groups of 3 negative chips, respectively. Therefore, the products are:

$$3(-3) = -9, 2(-3) = -6, 1(-3) = -3, \text{ and } 0(-3) = 0.$$

B. In part (A), each time the number multiplied by -3 decreases by 1, the product increases by 3. This pattern suggests that the products are:

$$(-1)(-3) = 3, (-2)(-3) = 6, \text{ and } (-3)(-3) = 9.$$

In part (B) of Example 2, notice that when -3 is multiplied by a negative number, the result is positive. So when two integers have the same sign, their product is positive. The box below summarizes the rules about the sign of the product of two integers.

Multiplying Integers

The product of two integers with the same sign is positive.
The product of two integers with different signs is negative.

EXAMPLE 3

Find the product $(-17)(30)$.

SOLUTION

The two integers have different signs, so the product is negative. Thus, just multiply the two integers as if they were both positive and then attach a negative sign to the result. $(-17)(30) = -510$

For Exercises 4–6, find each product.

4. $(-5)(12)$ **5.** $8(-4)$ **6.** $(-16)(6)$

7. Determine whether the product $3(-4)(-5)$ is positive or negative. Explain.

(continued)

NAME _____ DATE _____

Multiplying and Dividing Integers

To divide two integers, recall the relationship between multiplication and division.

$12 \div 3 = 4$ because $4(3) = 12$.

$-12 \div 3 = -4$ because $(-4)(3) = -12$.

$12 \div (-3) = -4$ because $(-4)(-3) = 12$.

$-12 \div (-3) = 4$ because $4(-3) = -12$.

The rules for dividing two integers are based on those for multiplying two integers.

Dividing Integers

The quotient of two integers with the same sign is positive.
The quotient of two integers with different signs is negative.

EXAMPLE 4

Find each quotient.

A. $36 \div (-9)$

B. $-36 \div (-9)$

SOLUTION

A. The signs are different, so the quotient is negative.

$36 \div (-9) = -4$

B. The signs are the same, so the quotient is positive.

$-36 \div (-9) = 4$

Find each quotient.

8. $-28 \div (-7)$　　　　**9.** $36 \div (-12)$　　　　**10.** $-64 \div 16$

Mixed Review

11. A credit card statement shows total charges of $124 and a credit of $150 for a returned item. If the balance last month was $215, what is the new balance for this month?

12. Write $\frac{7}{8}$ as a decimal and as a percent.

NAME _____ DATE _____

Quick Check

Review of Topic 5, Lesson 3

Standardized Testing Quick Check

1. Which statement is always true?

    A. A negative number minus a negative number is negative.

    B. A negative number plus a negative number is negative.

    C. A positive number plus a negative number is positive.

    D. A positive number minus a negative number is negative.

Homework Review Quick Check

State the opposite of each number.

2. -41 **3.** 23 **4.** 81

Find each difference.

5. $12 - 64$ **6.** $-31 - 22$ **7.** $17 - (-35)$

8. $-2 - 93$ **9.** $-51 - (-7)$ **10.** $-42 - (-42)$

Algebra 1
Basic Skills Workbook: Diagnosis and Remediation

Practice

For use with Lesson 5.4: Multiplying and Dividing Integers

For Exercises 1–45, find each product or quotient.

1. $12(-7)$

2. $64 \div (-4)$

3. $90 \div 5$

4. $-88 \div 11$

5. $(-24)(10)$

6. $98 \div (-7)$

7. $15(20)$

8. $30 \div 6$

9. $(-18)(-3)$

10. $-82 \div 2$

11. $-105 \div (-7)$

12. $400 \div (-20)$

13. $42(-3)$

14. $(-11)(-11)$

15. $-78 \div 6$

16. $26 \div (-2)$

17. $(-2)(52)$

18. $17(3)$

19. $22(-5)$

20. $75 \div (-3)$

21. $(-12)(-13)$

22. $(-8)(22)$

23. $(-21)(-4)$

24. $42 \div (-7)$

25. $-125 \div (-5)$

26. $(-5)(-16)$

27. $(-3)(33)$

28. $30(-9)$

29. $96 \div 2$

30. $-39 \div 13$

31. $96 \div (-24)$

32. $60(-9)$

33. $10(-13)$

34. $250 \div (-5)$

35. $68 \div (-4)$

36. $55 \div 5$

37. $(-25)(-40)$

38. $-48 \div 16$

39. $(-4)(-5)(-10)$

40. $96 \div (-4)$

41. $(-4)(-4)(-4)$

42. $-16 \div (-2)$

43. $(-32)(9)$

44. $2(-12)(-3)$

45. $56 \div (-7)$

46. A nurse checked a patient's temperature every hour for 3 h. Each time, the temperature had fallen 1°F. Express the change in the patient's temperature over that time period as an integer.

47. A football team lost 5 yd on each of three consecutive plays. Express the total yardage for the three plays as an integer.

48. Frank hopes to experience a weight change of -20 lb over the next 10 weeks. Use an integer to express the average weight change per week he hopes to achieve.

NAME _____ DATE _____

Assessment

For use with Topic 5: Integers

Name the opposite of the number.

1. -118 **2.** -54 **3.** 4

Find the absolute value.

4. $|-20|$ **5.** $|85|$ **6.** $|-60|$

Complete the statement using <, >, or =.

7. -9 ▢ 9 **8.** -22 ▢ 0 **9.** -30 ▢ -21

Find each sum.

10. $12 + 27$ **11.** $-15 + 41$ **12.** $-16 + (-14)$

13. $8 + (-26)$ **14.** $-30 + 27$ **15.** $-24 + (-3)$

Find each difference.

16. $13 - (-5)$ **17.** $-17 - 21$ **18.** $82 - 65$

19. $20 - (-3)$ **20.** $50 - 61$ **21.** $-9 - 3$

Find each product.

22. $15(80)$ **23.** $(-11)(-7)$ **24.** $12(-9)$

25. $(-30)(7)$ **26.** $(-9)(-41)$ **27.** $10(-19)$

Find each quotient.

28. $57 \div (-3)$ **29.** $-21 \div 3$ **30.** $65 \div 13$

31. $-84 \div (-6)$ **32.** $-19 \div (-1)$ **33.** $54 \div (-18)$

34. The lowest elevation in Death Valley National Park in California is -282 ft
(282 ft below sea level). The elevation of the Park Headquarters is -190 ft.
Find the difference in elevation between the two locations.

NAME _____ DATE _____

Cumulative Assessment

For use after Topics 1–5

The bar graph at the right shows the number of wins for the Major League Baseball teams that finished first in their divisions in 1996.

1. Find the mean, the median, the mode(s), and the range of the data.

2. Write the ratio of Indians' wins to Cardinals' wins as a fraction in lowest terms.

3. Could you draw a line graph to display the data? a circle graph? Explain.

4. Explain why the bar graph could be misleading.

Division Leader's Wins, 1996

Write each given number as a fraction or mixed number in lowest terms, then write it as a decimal and as a percent.

5. $\dfrac{78}{100}$

6. $\dfrac{21}{12}$

7. $\dfrac{4}{800}$

For Exercises 8–10, write each percent as a decimal and as a fraction or mixed number in lowest terms.

8. 0.8%

9. 96%

10. 325%

Find the missing number.

11. $\dfrac{12}{42} = \dfrac{?}{7}$

12. $\dfrac{21}{84} = \dfrac{?}{28}$

13. 25% of 88 is what number?

14. What number is 7% of 50?

Find the area and perimeter or circumference of each figure. In Exercise 17, use 3.14 for π.

15.

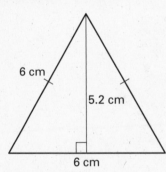

6 cm 6 cm

5.2 cm

6 cm

16.

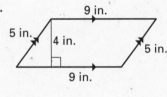

9 in.

5 in. 4 in. 5 in.

9 in.

17.

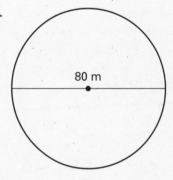

80 m

Perform each operation.

18. $153.6 + 97.8$

19. $72.4 - 13.75$

20. 10.35×2.68

21. $69.803 \div 8.3$

22. 16×3.87

23. $320 \div 1.6$

24. $-32 + 25$

25. $102 \div (-3)$

26. $(-81)(-4)$

27. $21 - 73$

28. $20(-13)$

29. $-33 + (-16)$

30. $-40 \div (-10)$

31. $-77 - 48$

32. $-52 - (-33)$

Algebra 1
Basic Skills Workbook: Diagnosis and Remediation

TOPICS

1–5

NAME _____ DATE _____

Cumulative Assessment

For use after Topics 1–5

Perform the operation.

33. $2\dfrac{3}{8} + 4\dfrac{6}{8}$

34. $\dfrac{8}{15} + \dfrac{2}{9}$

35. $5\dfrac{1}{6} - 1\dfrac{7}{10}$

36. $\dfrac{3}{4} \times 3\dfrac{2}{3}$

37. $6 \div \dfrac{3}{8}$

38. $\dfrac{4}{5} \div 2\dfrac{1}{7}$

Find the opposite and the absolute value of the integer.

39. -87

40. 0

41. 53

Write the numbers in order from least to greatest.

42. $3, -7, 0, -2, 5$

43. $2.4, 2.3999, 2.04, 20.4$

44. $\dfrac{3}{4}, \dfrac{4}{5}, \dfrac{7}{10}, \dfrac{13}{20}$

45. $1\dfrac{3}{4}, \dfrac{8}{3}, \dfrac{5}{2}, 2\dfrac{5}{8}, \dfrac{15}{8}$

108

Algebra 1
Basic Skills Workbook: Diagnosis and Remediation

Copyright © McDougal Littell Inc.
All rights reserved.

ANSWERS

Diagnostic Test *pages v–x*

1. 21:1, 3, 7, 21; 49:1, 7, 49; 7; 147

2. 99:1, 3, 9, 11, 33, 99; 33:1, 3, 11, 33; 33; 99

3. 48:1, 2, 3, 4, 6, 8, 12, 16, 24, 48; 52:1, 2, 4, 13, 26, 52; 4; 624 **4.** $2^2 \cdot 3 \cdot 7$

5. $3^2 \cdot 13$ **6.** prime **7.** > **8.** > **9.** =

10. 3168, 3367, 3368, 3370 **11.** 16.0009, 16.005, 16.01, 16.42 **12.** $\frac{3}{7}, \frac{1}{2}, \frac{2}{3}, \frac{8}{9}$

13. $3\frac{1}{2}, 3\frac{3}{4}, 4\frac{1}{3}, 4\frac{1}{2}$ **14.** 384 **15.** 288.67

16. 101.04 **17.** 35.26 **18.** 10.7 **19.** 15.36

20. $1\frac{1}{9}$ **21.** $\frac{1}{10}$ **22.** $4\frac{2}{3}$ **23.** $\frac{5}{3}$ **24.** $\frac{1}{16}$ **25.** $\frac{7}{45}$

26. 9 **27.** $\frac{7}{27}$ **28.** $3\frac{21}{22}$

29. $12.37; $12.99; $12.99; $6.00

30. 133.2; 126; None; 49 **31.** 50 cents/mi

32. Compact, Midsize, Minivan

33.

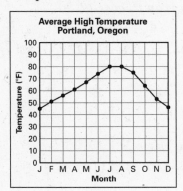

34. about 4.77 million **35.** about 1.35 million

36. graph B; no **37.** The break in the vertical scale of Graph B visually exaggerates the difference in height between the roller coaster in Japan and the other roller coasters. **38.** $\frac{1}{2}$ **39.** $\frac{6}{5}$

40. 55 mi/h **41.** $1500/month **42.** no **43.** 3

44. 36 **45.** 0.9 **46.** 1.625 **47.** $2.\overline{6}$ **48.** $\frac{1}{3}$

49. $6\frac{1}{3}$ **50.** $\frac{4}{7}$ **51.** $0.66\overline{6}; \frac{2}{3}$ **52.** $0.002; \frac{1}{500}$

53. 1.50; $1\frac{1}{2}$ **54.** 10 **55.** 30 **56.** 48 **57.** 9.9

58.

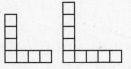

59. 13 squares

60. vertex **61.** quadrilateral **62.** square

63. Sketches may vary. An example is given.

64. 62 m; 150 m² **65.** 36 cm; 54 cm²

66. about 113.04 ft; about 1017.36 ft²

67. −16; 16 **68.** 8; 8 **69.** −3, 0, 1

70. −7, −2, 5, 6

71. 3 **72.** −66 **73.** −3 **74.** −45 **75.** −8

76. −16 **77.** −40 **78.** 8 **79.** 42

Topic 1
Lesson 1

Lesson Exercises *pages 2–3*

1. $2 \cdot 3^2$ **2.** $2 \cdot 3 \cdot 11$ **3.** $2^2 \cdot 3^2 \cdot 13$

4. $2 \cdot 5^2 \cdot 7$ **5.** 9 **6.** 3 **7.** 4 **8.** 1

9. 280 **10.** 576 **11.** 230 **12.** 42

13. *Sample answer:* about 48 **14.** 3600

Quick Check *page 4*

1. A **2.** A **3.** 6.2 **4.** no

Practice *page 5*

1. 1, 3, 5, 15 **2.** 1, 2, 13, 26

3. 1, 3, 9, 11, 33, 99 **4.** 1, 5, 7, 35

5. 1, 2, 3, 4, 6, 8, 9, 12, 16, 18, 24, 36, 48, 72, 144 **6.** 1, 7, 49 **7.** 1, 61 **8.** 1, 2, 4, 7, 8, 14, 28, 56 **9.** 1, 2, 3, 4, 6, 8, 9, 12, 18, 24, 36, 72

10. 1, 2, 7, 14, 49, 98 **11.** 1, 3, 7, 9, 21, 63

12. 1, 2, 4, 7, 14, 28, 49, 98, 196 **13.** 2^4

14. prime **15.** 3^2 **16.** 3^3 **17.** $2^2 \cdot 5$

18. $2 \cdot 3 \cdot 7$ **19.** 3^4 **20.** 2^8 **21.** $5 \cdot 13$

22. prime **23.** $5 \cdot 11$ **24.** $2^4 \cdot 3$

25. $2^3 \cdot 3 \cdot 5$ **26.** $2^2 \cdot 3^2 \cdot 7$ **27.** $7 \cdot 19$

28. prime **29.** 1 **30.** 1, 2, 4, 8

31. 1, 2, 3, 6 **32.** 1, 2, 3, 4, 6, 12 **33.** 1

34. 1, 17 **35.** 1, 2, 4, 8 **36.** 1, 3, 9

37. 3 **38.** 17 **39.** 8 **40.** 6 **41.** 1 **42.** 1

43. 2 **44.** 3 **45.** 1 **46.** 1 **47.** 2 **48.** 30

Answers

Topic 1 *continued*

49. 50 **50.** 2 **51.** 2 **52.** 7 **53.** 12 **54.** 72
55. 100 **56.** 45 **57.** 42 **58.** 208 **59.** 224
60. 1288 **61.** 51 **62.** 64 **63.** 144 **64.** 240
65. 300 **66.** 75 **67.** 30 **68.** 660

Lesson 2

Lesson Exercises *pages 6–8*

1. > **2.** > **3.** > **4.** < **5.** < **6.** >
7. 1304, 1334, 1340, 1430
8. 3.248, 3.284, 3.481, 3.847 **9.** $\frac{1}{3}, \frac{5}{8}, \frac{5}{6}, \frac{6}{5}$
10. $\frac{1}{3}, \frac{3}{8}, \frac{7}{12}, \frac{3}{4}$ **11.** > **12.** = **13.** <
14. 6 **15.** 8 **16.** 11 **17.** 1 **18.** 63
19. 140 **20.** 2125 **21.** 24

Quick Check *page 9*

1. B **2.** D **3.** 6 **4.** 8 **5.** 1 **6.** 147
7. 969 **8.** 36

Practice *page 10*

1. < **2.** > **3.** < **4.** < **5.** < **6.** > **7.** <
8. < **9.** > **10.** < **11.** = **12.** > **13.** =
14. > **15.** > **16.** > **17.** = **18.** < **19.** <
20. > **21.** < **22.** 5840, 8054, 8450, 8455
23. 20,599, 21,431, 23,546, 23,766
24. 12.05, 12.52, 21.25, 21.257
25. 7.0008, 7.0056, 7.0074, 7.059
26. $\frac{1}{6}, \frac{2}{3}, \frac{3}{4}, \frac{7}{8}$ **27.** $\frac{1}{4}, \frac{3}{10}, \frac{9}{20}, \frac{1}{2}, \frac{3}{5}$
28. $7\frac{1}{2}, 7\frac{2}{3}, 7\frac{7}{9}, 8\frac{1}{6}, 8\frac{5}{18}$ **29.** $1\frac{1}{10}, \frac{9}{7}, 1\frac{2}{5}, \frac{55}{35}$
30. $\frac{11}{9}, \frac{17}{11}, \frac{15}{9}, \frac{13}{3}$ **31.** $14\frac{1}{3}, 14\frac{13}{15}, 15\frac{1}{2}, 15\frac{4}{5}$
32. Mark **33.** no

Lesson 3

Lesson Exercises *pages 11–14*

1. 47 **2.** 830 **3.** 28,430 **4.** 614 **5.** 5734
6. 34,921 **7.** $2\frac{7}{12}$ **8.** $25\frac{3}{19}$ **9.** 14.11
10. 9.36 **11.** 4.45 **12.** 4.348 **13.** 12.6225
14. 9.931285 **15.** 0.102 **16.** 0.299268
17. 6.135 **18.** 19 **19.** 1308.33 **20.** 40.4
21. $2^5 \cdot 3$ **22.** prime **23.** $3 \cdot 5 \cdot 7$
24. $7 \cdot 13$ **25.** 5.23, 5.25, 5.79, 6.0
26. 6585, 8556, 8557, 8565 **27.** $\frac{1}{2}, \frac{5}{9}, \frac{4}{6}, \frac{5}{3}$
28. 179.0089, 179.009, 179.087, 179.090

Quick Check *page 15*

1. C **2.** A **3.** 5.4, 5.45, 6.75, 6.8, 7.05, 7.5
4. $\frac{7}{9}, \frac{8}{7}, \frac{7}{3}, \frac{18}{6}, \frac{13}{4}, \frac{7}{2}$ **5.** = **6.** = **7.** >

Practice *page 16*

1. 130 **2.** 366 **3.** 21 **4.** 58 **5.** 14.42
6. 21.36 **7.** 1.02 **8.** 14.95 **9.** 18.64
10. 6.181 **11.** 122.312 **12.** 0.49 **13.** 2.542
14. 36.924 **15.** 8.59 **16.** 876.29 **17.** 0.00015
18. 295.9858 **19.** 10.992 **20.** 5.697
21. 7.683 **22.** 40.625 **23.** 2.4908 **24.** 16.7
25. 5.6 **26.** 81 **27.** 52.4875 **28.** 19,100
29. 3.6 **30.** 18.4 **31.** 22.3 **32.** 578.6 **33.** 4
34. 31 **35.** 2.4625 **36.** 72.7824 **37.** 4.78
38. 4220 **39.** 224.64 **40.** 520.37908 **41.** 1.875
42. 77.49 **43.** $62.44 **44.** $4.03
45. 205.2 cm^2

Lesson 4

Lesson Exercises *pages 17–20*

1. $\frac{1}{2}$ **2.** $\frac{3}{4}$ **3.** $\frac{4}{7}$ **4.** $1\frac{1}{3}$ **5.** $\frac{17}{24}$ **6.** $\frac{1}{3}$ **7.** $9\frac{1}{4}$
8. $6\frac{1}{3}$ **9.** $\frac{2}{3}$ **10.** $\frac{39}{56}$ **11.** 33 **12.** $\frac{45}{64}$ **13.** $\frac{13}{7}$
14. $\frac{5}{17}$ **15.** $\frac{17}{46}$ **16.** $\frac{5}{22}$ **17.** $\frac{49}{25}$ **18.** 20
19. 12 **20.** > **21.** > **22.** =
23. 6 ft^2, or 864 in.^{2}

Quick Check *page 21*

1. B **2.** C **3.** 260 **4.** 21.05 **5.** 4404
6. 533.4

Practice *page 22*

1. $1\frac{1}{6}$ **2.** $\frac{1}{10}$ **3.** $\frac{1}{2}$ **4.** $\frac{3}{14}$ **5.** $\frac{8}{21}$ **6.** 1 **7.** $12\frac{4}{5}$
8. $2\frac{13}{24}$ **9.** $7\frac{4}{9}$ **10.** $11\frac{1}{5}$ **11.** $4\frac{1}{15}$ **12.** $2\frac{5}{16}$
13. 2 **14.** $\frac{1}{25}$ **15.** $\frac{8}{47}$ **16.** 6 **17.** $2\frac{2}{11}$ **18.** $\frac{15}{28}$
19. $\frac{2}{27}$ **20.** $\frac{3}{20}$ **21.** $\frac{1}{4}$ **22.** $\frac{6}{7}$ **23.** $6\frac{3}{16}$ **24.** $4\frac{7}{8}$
25. $5\frac{3}{4}$ **26.** $3\frac{3}{4}$ **27.** $13\frac{1}{5}$ **28.** $1\frac{1}{35}$ **29.** $\frac{5}{6}$
30. $24\frac{2}{3}$ **31.** 10 times **32.** 5 cups

Topic 1 Assessment *page 23*

1. 1; 702 **2.** 4; 32 **3.** 1; 217 **4.** 4; 1584
5. 15; 300 **6.** 18; 324 **7.** > **8.** = **9.** >
10. $\frac{1}{7}, \frac{4}{7}, \frac{3}{4}, \frac{3}{2}$ **11.** 0.0035, 0.0053, 0.053, 0.054
12. 2.345, 2.4, 2.61 **13.** 129 **14.** 98.989
15. 3 **16.** $1\frac{1}{18}$ **17.** $12\frac{1}{18}$ **18.** $3\frac{13}{20}$ **19.** 46

Topic 1 *continued*

20. 58.89 **21.** $\frac{2}{5}$ **22.** $\frac{5}{42}$ **23.** $3\frac{5}{6}$ **24.** $\frac{17}{48}$

25. 9675 **26.** 252.78 **27.** $\frac{2}{3}$ **28.** 1948.544

29. $\frac{20}{143}$ **30.** 19.7967 **31.** $37\frac{4}{5}$ **32.** 341.1

33. $1\frac{3}{4}$ **34.** 20 **35.** $6\frac{3}{4}$ **36.** 2.34

37. 33.924 lb **38.** $1\frac{3}{8}$ miles

Topic 2
Lesson 1

Lesson Exercises *pages 25–26*

1. 0.85 **2.** 97 **3.** 100.2 **4.** 7.25 **5.** 142

6. 15.75 **7.** 24 **8.** 70 **9.** no mode

10. 35 and 37 **11.** 4 **12.** 42 **13.** 45 **14.** $6\frac{1}{4}$

15. 700 **16.** 8.15 **17.** 561.07 **18.** $28\frac{7}{8}$

Quick Check *page 27*

1. B **2.** D **3.** $\frac{51}{56}$ **4.** $\frac{7}{20}$

Practice *page 28*

For Exercises 1–8, the answers are given in the order: mean, median, mode(s), and range.

1. about 21.9; 20; 16, 17, and 32; 26

2. about 7.5; about 7.5; no mode; about 1.1

3. 28; 24; 18; 36 **4.** $23.97; $27.45; 0; $49.95

5. about 109.3; 100; 88 and 100; 72

6. about 72.3°F; 70°F; 70°F; 18°F

7. about 77.7; 66; no mode; 105

8. 787.5; 767.6; no mode; 552.8

Lesson 2

Lesson Exercises *pages 30–32*

1. Answers may vary. *Sample answer:* about 43¢/mi

2. The operating cost for a full-size car is about 10¢/mi greater than for a compact.

3. Answers may vary. An example is given. about 21.1 mi/gal

4. It increased; about 0.2 mi/gal.

For Exercises 5 and 6, graphs may vary. Examples are given.

5.
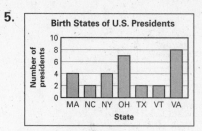

6.
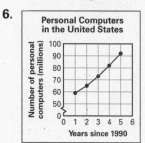

7. Answers may vary. The data can be displayed in a bar graph because the years can be thought of as categories.

8. mean: about 4.1; median: 4; mode: 2; range: 6

9. lcm: 140, gcf: 7 **10.** 1, 2, 4, 8, 16

Quick Check *page 33*

1. C **2.** B

For Exercises 3 and 4, the answers are given in the order: mean, median, mode(s), range.

3. 26.4; 12; no mode; 64 **4.** 32.5; 32; 33; 35

Practice *page 34*

For Exercises 1–3, answers may vary. Examples are given.

1. about 4.5 calories **2.** about 5.5 calories

3. about 6.5 calories

4. jogging, downhill skiing, and tennis

5. tennis

For Exercises 6–8, answers may vary. Examples are given.

6. about 105 calories **7.** about 29

8. about 39 **9.** 1993 **10.** 1995 **11.** about 20

For Exercises 12 and 13, graphs may vary. Examples are given.

Answers

Topic 2 *continued*

12.

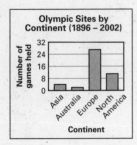

13.

Lesson 3

Lesson Exercises *pages 36–37*

1. about 13.3 quadrillion Btu's

2. about 53.9 quadrillion Btu's

3. about 7.7%   4. about 7.9%

5. about 84.5%   6. 83   7. 52   8. $\frac{1}{10}$

9. 48   10. 57.242   11. 3.5

Quick Check *page 38*

1. C

For Exercises 2–5, answers may vary. Examples are given.

2. about 55   3. about 60   4. about 200 million

5. about 100 million

Practice *page 39*

1. about 30,000 persons

2. about 27,000 persons

3. about 24,000 persons

4. about 19,000 persons

5. about 62,500 persons

6. about 127,720,000

7. about 74,160,000

8. about 4,120,000

9. about 8,137,000 registered motor vehicles

10. 140 ski runs

---

11. **a.** 30% **b.** 45% **c.** 25%

12. Answers may vary. An example is given. I think the greatest number of skiers probably ski at the intermediate level.

Lesson 4

Lesson Exercises *pages 41–42*

1. Graph B gives the impression that Ride 3 is twice as steep as Ride 5.

For Exercises 2–4, answers may vary. Examples are given.

2. Graph A gives a more favorable impression of the steepness of Ride 3 in relation to the steeper roller coasters.

3. A candidate running for reelection might want to create the impression that the median income has improved sharply during his or her current term in office.

4. The graph would look even flatter and give the impression of a slower or smaller increase overall.

5. 20   6. Germany   7. 2

Quick Check *page 43*

1. A   2. 74%   3. about 51,570,000 acres

Practice *page 44*

For Exercises 1 and 2, answers may vary. Examples are given.

1. The graph makes it appear that the gourmet coffee at Cafe Break is more than three times as expensive as the gourmet coffee at Alonzo's.

2. The prices appear to vary widely, when the range is only about 75¢.

3. Alonzo's; Cafe Break

For Exercises 4 and 5, answers may vary. Examples are given.

4. Make the vertical scale range from $0 to $7.25.

5. Graph A makes it appear that the median size of supermarkets increased sharply from 1991 to 1995, while Graph B makes the increase appear smaller.

6. The developer should use Graph A to convince the investors that they need a larger store to compete in a market where stores are rapidly getting larger.

7. false   8. false   9. true

Topic 2 *continued*

Topic 2 Assessment *page 45*

1. $7\frac{1}{3}$; 6; no mode; 13 **2.** 30.2; 30; 30; 4

3. about 2.5 million dollars **4.** Martinez **5.** 3 players

6. Answers may vary. An example is given. about 71 years

7. 1940 to 1950

8. bar graph; Graphs may vary. An example is given.

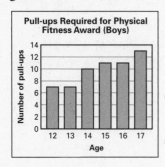

9. $7.5 billion

10. The scale begins at 0 and continues without a break, so the lengths of the bars can actually be compared directly.

11. Answers may vary. An example is given. Start the scale at $0.5 million, with intervals of $0.25 million.

Topic 3
Lesson 1

Lesson Exercises *pages 47–48*

For Exercises 1–6, answers are given in fraction form.

1. $\frac{13}{3}$ **2.** $\frac{3}{2}$ **3.** $\frac{2}{1}$ **4.** $\frac{50}{3}$ **5.** $\frac{21}{8}$

6. $\frac{18}{1}$ **7.** $5 per lb **8.** 300 mi per day

9. $290 per week **10.** 13 m per s

11. Because the vertical scale does not begin at 0, the increase appears dramatic when it is actually very slight.

12. 8; 792 **13.** $5.47

Quick Check *page 49*

1. C **2.** false **3.** true **4.** false

Practice *page 50*

For Exercises 1–15, answers are in fraction form.

1. $\frac{4}{25}$ **2.** $\frac{87}{4}$ **3.** $\frac{7}{1}$ **4.** $\frac{73}{29}$ **5.** $\frac{17}{49}$ **6.** $\frac{9}{5}$

7. $\frac{3}{2}$ **8.** $\frac{56}{1}$ **9.** $\frac{24}{1}$ **10.** $\frac{500}{1}$ **11.** $\frac{9}{1}$ **12.** $\frac{12}{1}$

13. $\frac{24}{5}$ **14.** $\frac{16}{1}$ **15.** $\frac{2}{1}$ **16.** 45 mi/h

17. $8.50/h **18.** 2 lb/week **19.** $7.50/ticket

20. $2.75/lb **21.** 60 words/min **22.** 7.5 m/s

23. $150/month **24.** $1.40/gal **25.** 90 km/h

26. 45 revolutions/min **27.** 25 students/teacher

Lesson 2

Lesson Exercises *page 52*

1. yes; $\dfrac{1}{3} = \dfrac{1 \cdot 4}{3 \cdot 4} = \dfrac{4}{12}$

2. no; $\dfrac{2}{7} = \dfrac{2 \cdot 8}{7 \cdot 8} = \dfrac{16}{56}$

3. yes; $\dfrac{5}{6} = \dfrac{5 \cdot 5}{6 \cdot 5} = \dfrac{25}{30}$

4. yes; $\dfrac{7}{8} = \dfrac{7 \cdot 3}{8 \cdot 3} = \dfrac{21}{24}$ **5.** no **6.** Numbers increase by 3; 18, 21, 24 **7.** 320 calories

Quick Check *page 53*

1. B **2.** B **3.** 11 : 9 **4.** 2 to 3 **5.** 8 to 7

6. 48.6 h/year; about 32 movies/year

Practice *page 54*

1–4. Equivalent fractions may vary.

1. D; $\frac{12}{18}$ **2.** B; $\frac{6}{8}$ **3.** A; $\frac{4}{16}$ **4.** C; $\frac{6}{12}$ **5.** 18

6. 40 **7.** 6 **8.** 9

9–12. Sample answers are given. Check explanations.

9. $\frac{12}{22}, \frac{24}{44}, \frac{30}{55}$ **10.** $\frac{8}{18}, \frac{12}{27}, \frac{24}{54}$ **11.** $\frac{1}{2}, \frac{3}{6}, \frac{14}{28}$

12. $\frac{10}{32}, \frac{15}{48}, \frac{20}{64}$ **13.** true **14.** false; 60

15. false; 4 **16.** true **17.** 10 cm **18.** 3 feet

19. yes; $\dfrac{10 \text{ mi}}{75 \text{ min}} = \dfrac{2 \text{ mi}}{15 \text{ min}}$ and $\dfrac{4 \text{ mi}}{30 \text{ min}} = \dfrac{2 \text{ mi}}{15 \text{ min}}$

20. 4 cm

Lesson 3

Lesson Exercises *pages 56–58*

1. $0.41\overline{6}$ **2.** $1.\overline{7}$ **3.** 3.8 **4.** $\frac{1}{4}$ **5.** $\frac{3}{5}$ **6.** $3\frac{2}{3}$

7. 85% **8.** $41.\overline{6}$% **9.** 180% **10.** 0.65; $\frac{13}{20}$

Answers

Topic 3 *continued*

11. 2.2; $2\frac{1}{5}$ **12.** 0.005; $\frac{1}{200}$ **13.** 50%, $\frac{1}{2}$

14. 18.5%, $\frac{37}{200}$ **15.** 132%, $1\frac{8}{25}$ **16.** 12

17. circle

Quick Check *page 59*

1. C **2.** A **3.** no; $\frac{2 \cdot 3}{3 \cdot 3} = \frac{6}{9}$ **4.** 7

Practice *page 60*

1. 0.75 **2.** 0.7 **3.** $0.\overline{3}$ **4.** $0.4\overline{3}$ **5.** 8.25

6. 0.5625 **7.** 5.74 **8.** $0.\overline{27}$ **9.** $6\frac{1}{2}$ **10.** $\frac{1}{6}$

11. $\frac{5}{9}$ **12.** $\frac{18}{25}$ **13.** $\frac{1}{2}$ **14.** $5\frac{4}{5}$ **15.** $7\frac{2}{13}$ **16.** $2\frac{2}{3}$

17. 80% **18.** 12.5% **19.** 85% **20.** 250%

21. 90% **22.** 575% **23.** 90% **24.** $33.\overline{3}\%$

25. 37.5% **26.** 140% **27.** 204% **28.** 798%

29. 0.51 **30.** $0.52\overline{3}$ **31.** 1.02 **32.** 0.025

33. 0.0075 **34.** 0.001 **35.** 0.09 **36.** 2.34

37. $\frac{99}{100}$ **38.** $1\frac{1}{4}$ **39.** $\frac{1}{2}$ **40.** $\frac{3}{100}$ **41.** $\frac{3}{400}$ **42.** $\frac{1}{8}$

43. $1\frac{1}{2}$ **44.** $\frac{2}{25}$ **45.** $\frac{1}{50}$ **46.** $\frac{1}{25}$ **47.** $2\frac{1}{4}$ **48.** $1\frac{1}{5}$

49. 25%, $\frac{1}{4}$ **50.** 67.5%, $\frac{27}{40}$ **51.** 120%, $1\frac{1}{5}$

52. 12.8%, $\frac{16}{125}$ **53.** 6%, $\frac{3}{50}$ **54.** 37.5%, $\frac{3}{8}$

55. 76%, $\frac{19}{25}$ **56.** 560%, $5\frac{3}{5}$ **57.** 44%, $\frac{11}{25}$

58. 475%, $4\frac{3}{4}$ **59.** 0.1%, $\frac{1}{1000}$ **60.** 101%, $1\frac{1}{100}$

Lesson 4

Lesson Exercises *pages 61–62*

1. 137 **2.** 627 **3.** 168.42 **4.** 204.204

5. 0.615 **6.** 91.3936 **7.** $26.99, $152.91

8. $8.03, $160.94 **9.** 25,600 **10.** 21,600

11. 3680 **12.** $\frac{1}{240}$ **13.** 0.96, 96%

Quick Check *page 63*

1. B **2.** B **3.** $1\frac{3}{4}$ **4.** $\frac{23}{50}$ **5.** $\frac{4}{25}$ **6.** $\frac{7}{25}$

7. $36.\overline{36}\%$ **8.** $108.\overline{3}\%$ **9.** 680% **10.** $11.\overline{1}\%$

Practice *page 64*

1. 24 **2.** 6 **3.** 36 **4.** 255 **5.** 68.75

6. 50.4 **7.** 19.6 **8.** 66.96 **9.** 124.5

10. 167.32 **11.** 1128.6 **12.** 0.095

13. 6.4975 **14.** 2169.42 **15.** 158.355

16. 546.112 **17.** $1.30 **18.** $3.28, $13.12

19. about 49 **20.** $6.21 **21.** $7.83

22. Urban: 138.7395 mi^2, Rural: 9110.5605 mi^2

23. Urban: 8110.6064 mi^2, Rural: 147,862.5936 mi^2

24. Urban: 209.1422 mi^2, Rural: 1745.4578 mi^2

25. Urban: 2425.9476 mi^2, Rural: 4992.8524 mi^2

26. Urban: 988.2495 mi^2, Rural: 108,817.2505 mi^2

Topic 3 Assessment *page 65*

1. 0.86; 86% **2.** 3.6; 360% **3.** 0.375; 37.5%

4. $\frac{3}{5}$ **5.** $\frac{3}{7}$ **6.** $7\frac{1}{5}$ **7.** 0.96; $\frac{24}{25}$ **8.** 1.1; $1\frac{1}{10}$

9. 0.06; $\frac{3}{50}$ **10.** $\frac{25}{24}$ **11.** $\frac{24}{5}$ **12.** $\frac{36}{5}$ **13.** $\frac{7}{3}$

14. $\frac{15}{1}$ **15.** 24 points/game **16.** 27.5 mi/gal

17. $3.25/lb **18.** 47 pages/chapter **19.** 16

20. 3 **21.** 4 **22.** 9 **23.** 14.4 **24.** 0.4

25. 24.5 **26.** 192

Topic 4
Lesson 1

Lesson Exercises *page 67*

1. C **2.**

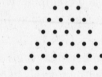

$$\frac{8(8 + 1)}{2} = 4(9) = 36$$

3. 120; 210; 465

4. 139,870,000 square miles (or about 140 million square miles)

5. $4.50/ticket

Quick Check *page 68*

1. D **2.** B **3.** 17 **4.** 1.35 **5.** 12 dentists

Practice *page 69*

1. 6; 12; 20 **2.** 30, 42, and 56 **3.** $n(n + 1)$

4. 240 **5.** 2; 3; 4

6. The number of triangles is 2 less than the number of sides.

7.

5 triangles   6 triangles

Topic 4 *continued*

8. $n - 2$ **9.** 18 triangles

10.

11. 3; 6; 10; 15; 21 **12.** $1 + 2 + 3 + \ldots +$
$(n - 1)$, or $\dfrac{n(n - 1)}{2}$

Lesson 2

Lesson Exercises *page 72*

1. a. The figure is not closed.

    b. Answers may vary. An example is given. It
has line segments that intersect more than
two sides.

2. The polygon is a pentagon. Names may vary.
Two examples are *VWXYZ* and *WVZYX*; sides $\overline{VW}$,
$\overline{WX}$, $\overline{XY}$, $\overline{YZ}$, and $\overline{VZ}$; vertices: *V*, *W*, *X*, *Y*, and *Z*;
diagonals: $\overline{VX}$, $\overline{VY}$, $\overline{WZ}$, $\overline{WY}$, and $\overline{XZ}$.

3. Answers may vary. An example is given.

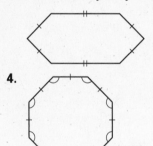

4.

5. 7.5 **6.** $3n$

Quick Check *page 73*

1. B **2.** D **3.** $2n + 1$

Practice *page 74*

1. yes **2.** yes

3. No; the figure is not formed entirely by line
segments. **4.** five **5.** congruent **6.** regular

7. parallelogram **8.** quadrilateral **9.** rectangle

For Exercises 10–15, the names of the polygon
may vary. Examples are given.

10. *JKLMNP*, *MNPJKL*, and *JPNMLK*; $\overline{JK}$, $\overline{KL}$,
$\overline{LM}$, $\overline{MN}$, $\overline{PN}$, and $\overline{JP}$; *J*, *K*, *L*, *M*, *N*, and *P*; $\overline{JN}$,
$\overline{JM}$, $\overline{JL}$, $\overline{KP}$, $\overline{KN}$, $\overline{KM}$, $\overline{LN}$, $\overline{LP}$, and $\overline{MP}$

11. *DEFG*, *FGDE*, and *EDGF*; $\overline{DE}$, $\overline{EF}$, $\overline{GF}$, and
$\overline{DG}$; *D*, *E*, *F*, and *G*; $\overline{DF}$ and $\overline{GE}$

12. *JKL*, *JLK*, and *KJL*; $\overline{JK}$, $\overline{KL}$, and $\overline{JL}$; *J*, *K*, and
L; no diagonals

13. *ABCD*, *CDAB*, and *DABC*; $\overline{AB}$, $\overline{BC}$, $\overline{CD}$, and
$\overline{AD}$; *A*, *B*, *C*, and *D*; $\overline{AC}$ and $\overline{BD}$

14. *STUVWXYZ*, *VUTSZYXW*, and *ZSTUVWXY*;
$\overline{ST}$, $\overline{TU}$, $\overline{UV}$, $\overline{VW}$, $\overline{WX}$, $\overline{XY}$, $\overline{YZ}$, and $\overline{SZ}$; *S*, *T*, *U*, *V*,
W, *X*, *Y*, and *Z*; $\overline{SU}$, $\overline{SV}$, $\overline{SW}$, $\overline{SX}$, $\overline{SY}$, $\overline{TV}$, $\overline{TW}$, $\overline{TX}$,
$\overline{TY}$, $\overline{TZ}$, $\overline{UW}$, $\overline{UX}$, $\overline{UY}$, $\overline{UZ}$, $\overline{VX}$, $\overline{VY}$, $\overline{VZ}$, $\overline{WY}$, $\overline{WZ}$,
and $\overline{XZ}$

15. *PQRST*, *PTSRQ*, and *RSTPQ*; $\overline{PQ}$, $\overline{QR}$, $\overline{RS}$,
$\overline{ST}$, and $\overline{PT}$; *P*, *Q*, *R*, *S*, and *T*; $\overline{PR}$, $\overline{PS}$, $\overline{QS}$, $\overline{QT}$,
and $\overline{RT}$

For Exercises 16 and 17, answers may vary.
Examples are given.

16. **17.**

18. not possible; A square has four right angles.
Every square is a rectangle.

For Exercises 19–21, answers may vary. Examples
are given.

19. **20.**

21.

Lesson 3

Lesson Exercises *page 77*

1. 48 ft^2; 32 ft **2.** 81 ft^2; 36 ft **3.** 120 m^2; 46 m

4. 70 in.^{2}; 34 in. **5.** 3 in. and 4 in. **6.** $\frac{1}{90}$

Quick Check *page 78*

1. C **2. a.** pentagon **b.** no

    c. Answers may vary. Examples are
given. *GHIJK* and *GKJIH*

    d. $\overline{GH}$, $\overline{HI}$, $\overline{IJ}$, $\overline{JK}$, and $\overline{GK}$

    e. *G*, *H*, *I*, *J*, and *K*

    f. $\overline{GI}$, $\overline{GJ}$, $\overline{HJ}$, $\overline{HK}$, and $\overline{IK}$

3. Check sketches. The figure will be a square.

Answers

Topic 4 *continued*

Practice *page 79*

1. 78 ft   2. 26.4 in.   3. 114 cm   4. 30 in.

5. 28 cm   6. 31 m   7. 156.25 in.²   8. $7\frac{1}{2}$ yd²

9. 70 in.²   10. 157.5 km²   11. 150 mm²

12. 80 m²   13. 78 cm   14. 162 m²   15. 14 yd²

16. 120 ft; 900 ft²

Lesson 4

Lesson Exercises *pages 81–82*

1. about 25.12 cm   2. about 176 in.

3. about 78.5 yd²   4. about 154 m²

5. 15,000 ft²   6. They are the same.

7. **a.** CD sales   **b.** $77,000

Quick Check *page 83*

1. C   2. 18 ft   3. 32.5 cm   4. 600 m²

5. 48 in.²

Practice *page 84*

Answers found using 3.14 for π.

1. about 125.6 m   2. about 257.48 ft

3. about 175.84 cm   4. about 75.36 in.

5. about 18.84 yd   6. about 376.8 m

7. about 879.2 mm   8. about 314 m

9. about 163.28 ft   10. about 219.8 ft

11. about 153.86 cm²   12. about 7.07 ft²

13. about 314 in.²   14. about 346.19 yd²

15. about 38.47 m²   16. about 28.26 cm²

17. about 63.59 ft²   18. about 200.96 m²

19. about 254.34 yd²   20. about 15,386 in.²

21. about 42.04 ft; about 84.08 ft

22. about 10 m; about 20 m

23. about 31.4 ft   24. about 13.76 cm²

Topic 4 Assessment *pages 85–86*

1. figure 1: 3 units; figure 2: 4 units; figure 3: 5 units; figure 4: 6 units

2. $(n + 2)$ units   3. 22 units

4–6. Answers may vary. Examples are given.

4. 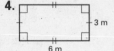   5.   6.

7. false   8. true   9. false   10. true

11. false   12. 46 cm   13. 15 m   14. 80 in.

15. 157.5 ft²   16. 720 mm²   17. 1008 yd²

18. about 314 cm   19. about 69.08 in.

20. about 21.98 yd   21. about 907.46 ft²

22. about 1256 m²   23. about 15,386 mm²

Topic 5

Lesson 1

Lesson Exercises *pages 88–89*

1. −6, −5, −1, 0, 3, 4

2. −10, −9, −8, −6, −4, −3

3. −25, −20, −18, 10, 16, 30

4.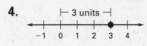

$|3| = 3$

5.
$|-1| = 1$

6.
$|0| = 0$

7.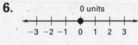

$|-15| = 15$

8. polygon: hexagon;

sides: $\overline{AB}, \overline{BC}, \overline{CD}, \overline{DE}, \overline{EF}, \overline{FA}$

vertices: A, B, C, D, E, F

diagonals: $\overline{AC}, \overline{AD}, \overline{AE}, \overline{BD}, \overline{BE}, \overline{BF},$
$\overline{CE}, \overline{CF}, \overline{DF}$

Quick Check *page 90*

1. A   2. C   3. about 12,462.66 m²

4. about 12.56 in.²   5. 31

6. about 17,662.5 mi²

Practice *page 91*

1.
2.
3.
4.

Algebra 1
Basic Skills Workbook: Diagnosis and Remediation

Topic 5 *continued*

5. (number line: $-8\ -6\ -4\ -2\ 0\ 2\ 4\ 6\ 8$)

6. (number line: $-20\ -16\ -12\ -8\ -4\ 0$)

7. 9 **8.** -15 **9.** -28 **10.** 4 **11.** 17

12. -100 **13.** $-46, -31, -22, 15, 25, 36$

14. $-110, -105, -103, -101, -100$

15. $30, 55, 60, 77, 85, 100$

16. $-25, -10, -3, 39, 51, 66$

17. $-110, -106, 106, 110, 116$

18. $-1110, -80, -79, 32, 85$

19. 21 **20.** 11 **21.** 34

22. 34 **23.** 15 **24.** 4 **25.** 19 **26.** 0

27. 13 **28.** = **29.** < **30.** = **31.** >

32. > **33.** < **34.** B **35.** G **36.** C

37. E **38.** D and F

Lesson 2

Lesson Exercises *pages 92–93*

1. 5 **2.** -5 **3.** -6 **4.** 30 **5.** -33

6. -20 **7.** -8 **8.** -9 **9.** -6

10. about 254.34 in.^{2} **11.** 54 km; 162 km^2

12. 170

Quick Check *page 94*

1. D **2.** C **3.** 6 **4.** 0 **5.** 17 **6.** 35

Practice *page 95*

1. never **2.** sometimes **3.** always

4. -10 **5.** 10 **6.** -2 **7.** 3 **8.** -1

9. -1 **10.** 8 **11.** 2 **12.** -82 **13.** -17

14. 87 **15.** -52 **16.** -14 **17.** -13

18. 144 **19.** 100 **20.** 25 **21.** -78 **22.** 11

23. -64 **24.** 7 **25.** 18 **26.** -45 **27.** 125

28. 8 **29.** 32 **30.** 13 **31.** 0 **32.** 48

33. 40 **34.** 105 **35.** -19 **36.** 68 **37.** 6

38. -81 **39.** 0 **40.** 16 **41.** -53 **42.** 104

43. 11°F **44.** a loss of 7 yards

Lesson 3

Lesson Exercises *pages 96–98*

1. -6 **2.** -4 **3.** 4 **4.** -1 **5.** 3 **6.** 5

7. 9 **8.** -21 **9.** 1 **10.** $8 + (-10); -2$

11. $12 + 7; 19$ **12.** $-13 + (-4); -!7$

13. $-4, -10$ **14.** a line graph **15.** 90 cm^2

16. $8

Quick Check *page 99*

1. B **2.** -50 **3.** 35 **4.** -7

Practice *page 100*

1. -35 **2.** 44 **3.** 0 **4.** 18 **5.** -71 **6.** 5

7. $1 + 15$ **8.** $51 + (-16)$ **9.** $32 + 19$

10. $8 + (-13)$ **11.** $12 + 17$ **12.** $2 + (-10)$

13. -19 **14.** -3 **15.** 3 **16.** 8 **17.** -8

18. 40 **19.** -30 **20.** 36 **21.** -17

22. -28 **23.** 65 **24.** 12 **25.** 18 **26.** 30

27. -62 **28.** -24 **29.** -13 **30.** -11

31. -6 **32.** -10 **33.** -41 **34.** 76

35. -30 **36.** 26 **37.** -103 **38.** -5

39. -16 **40.** -22 **41.** 16 **42.** -108

43. 39 **44.** 13 **45.** 10 **46.** 48 **47.** -9

48. -16 **49. a.** -8 ft (8 ft below sea level)

b. 274 ft lower

Lesson 4

Lesson Exercises *pages 101–103*

1. -10 **2.** -9 **3.** -8 **4.** -60 **5.** -32

6. -96

7. positive; Answers may vary. An example is given. The product $(-4)(-5)$ is positive, so $3(-4)(-5)$ is positive.

8. 4 **9.** -3 **10.** -4 **11.** $189

12. 0.875; 87.5%

Quick Check *page 104*

1. B **2.** 41 **3.** -23 **4.** -81 **5.** -52

6. -53 **7.** 52 **8.** -95 **9.** -44 **10.** 0

Practice *page 105*

1. -84 **2.** -16 **3.** 18 **4.** -8 **5.** -240

6. -14 **7.** 300 **8.** 5 **9.** 54 **10.** -41

11. 15 **12.** -20 **13.** -126 **14.** 121

15. -13 **16.** -13 **17.** -104 **18.** 51

19. -110 **20.** -25 **21.** 156 **22.** -176

Topic 5 *continued*

23. 84 **24.** −6 **25.** 25 **26.** 80 **27.** −99
28. −270 **29.** 48 **30.** −3 **31.** −4
32. −540 **33.** −130 **34.** −50 **35.** −17
36. 11 **37.** 1000 **38.** −3 **39.** −200
40. −24 **41.** −64 **42.** 8 **43.** −288
44. 72 **45.** −8 **46.** −3°F **47.** −15 yd
48. −2 lb/week

Topic 5 Assessment *page 106*

1. 118 **2.** 54 **3.** −4 **4.** 20 **5.** 85 **6.** 60
7. < **8.** < **9.** < **10.** 39 **11.** 26
12. −30 **13.** −18 **14.** −3 **15.** −27
16. 18 **17.** −38 **18.** 17 **19.** 23 **20.** −11
21. −12 **22.** 1200 **23.** 77 **24.** −108
25. −210 **26.** 369 **27.** −190 **28.** −19
29. −7 **30.** 5 **31.** 14 **32.** 19 **33.** −3
34. 92 ft

Cumulative Assessment *pages 107–108*

1. mean: about 92.7; median: 91.5; mode: none; range: 11

2. $\frac{9}{8}$ **3.** no; no; The data show neither change over time nor parts of a whole.

4. Because there is a break in the vertical scale, the relative lengths of the bars are misleading. For example, it appears the Yankees had three times as many wins as the Cardinals when they actually had only 4 more wins.

5. $\frac{39}{50}$; 0.78; 78% **6.** $1\frac{3}{4}$; 1.75; 175%

7. $\frac{1}{200}$; 0.005; 0.5% **8.** 0.008; $\frac{1}{125}$ **9.** 0.96; $\frac{24}{25}$

10. 3.25; $3\frac{1}{4}$ **11.** 2 **12.** 7 **13.** 22

14. 3.5 **15.** 15.6 cm²; 18 cm **16.** 36 in.²; 28 in.

17. about 5024 m²; about 251.2 m **18.** 251.4

19. 58.65 **20.** 27.738 **21.** 8.41 **22.** 61.92

23. 200 **24.** −7 **25.** −34 **26.** 324

27. −52 **28.** −260 **29.** −49

30. 4 **31.** −125 **32.** −19

33. $7\frac{1}{8}$ **34.** $\frac{34}{45}$ **35.** $3\frac{7}{15}$ **36.** $2\frac{3}{4}$ **37.** 16

38. $\frac{28}{75}$ **39.** 87; 87 **40.** 0; 0 **41.** −53; 53

42. −7, −2, 0, 3, 5 **43.** 2.04, 2.3999, 2.4, 20.4

44. $\frac{13}{20}, \frac{7}{10}, \frac{3}{4}, \frac{4}{5}$ **45.** $1\frac{3}{4}, \frac{15}{8}, \frac{5}{2}, 2\frac{5}{8}, \frac{8}{3}$

Answers